OPTICAL PROPERTIES OF SEMICONDUCTORS

OPTICHESKIE SVOISTVA POLUPROVODNIKOV

ОПТИЧЕСКИЕ   СВОЙСТВА   ПОЛУПРОВОДНИКОВ

The Lebedev Physics Institute Series

Editors: Academicians D. V. Skobel'tsyn and N. G. Basov

P. N. Lebedev Physics Institute, Academy of Sciences of the USSR

Recent Volumes in this Series

Proceedings (Trudy) of the P. N. Lebedev Physics Institute

Volume 75

OPTICAL PROPERTIES OF SEMICONDUCTORS

Edited by

N. G. Basov

P. N. Lebedev Physics Institute
Academy of Sciences of the USSR
Moscow, USSR

Translated from Russian by
Albin Tybulewicz

Editor: *Soviet Physics–Semiconductors*

CONSULTANTS BUREAU
NEW YORK AND LONDON

Library of Congress Cataloging in Publication Data

Main entry under title:

Optical properties of semiconductors.

(Proceedings (Trudy) of the P. N. Lebedev Physics Institute; v. 75)
Translation of Opticheskie svoĭstva poluprovodnikov.
Half title also in Russian.
"A special research report."
Includes bibliographical references.
1. Semiconductors–Optical properties–Addresses, essays, lectures. 2. Excitons–
Addresses, essays, lectures. I. Basov, Nikolaĭ Gennadievich, 1922-   II. Series:
Akademiĭa nauk SSSR. Fizicheskiĭ institut. Proceedings; v. 75.
QC1.A4114 vol. 75 [QC611.6.06]   530'.08s [537.6'22]

75-37609

The original Russian text was published by Nauka Press in Moscow in 1974 for the
Academy of Sciences of the USSR as Volume 75 of the Proceedings of the P. N.
Lebedev Physics Institute. This translation is published under an agreement with the
Copyright Agency of the USSR (VAAP).

ISBN 978-1-4615-7550-4   ISBN 978-1-4615-7548-1 (eBook)
DOI 10.1007/978-1-4615-7548-1

© 1976 Consultants Bureau, New York

A Division of Plenum Publishing Corporation
227 West 17th Street, New York, N. Y. 10011

United Kingdom edition published by Consultants Bureau, London
A Division of Plenum Publishing Company, Ltd..
Davis House (4th Floor), 8 Scrubs Lane, Harlesden, London, NW10 6SE, England

CONTENTS

Investigation of the Collective Properties of
Excitons in Germanium by Long-Wavelength
Infrared Spectroscopy Methods

V. A. Zayats

RADIATION EMITTED FROM SEMICONDUCTOR LASERS IN STRONG MAGNETIC FIELDS AND UNDER HIGH HYDROSTATIC PRESSURES*

I. I. Zasavitskii

An investigation was made of the influence of magnetic fields on the emission spectra of InAs, InSb, and PbSe injection lasers and of the influence of pressure on the tuning of the emission frequency of GaAs and PbSe lasers. Magnetically tunable stimulated Raman emission was obtained from n-type InSb crystals as a result of inelastic scattering of light ($\lambda = 10.6\ \mu$) accompanied by electron-spin flip. The effective masses and g factors of the carriers were determined. The dependences of the forbidden band width and refractive index on the applied pressure and magnetic field were obtained.

INTRODUCTION

Investigations of the energy structure of semiconductors are very desirable not only from the point of view of fundamental knowledge but also because of practical applications. Very interesting results are obtained when the energy spectrum and, therefore, electrical, optical, and other properties of a semiconductor are varied by external agencies such as temperature, pressure, or magnetic field.

The use of magnetic fields in measurements of the electrical and optical properties has given the most reliable information on the forbidden-band width, effective mass, and g factor of carriers, anisotropy of these quantities, and band nonparabolicity of many materials. Strong magnetic fields change radically the energy spectrum of a crystal because they quantize the allowed bands into Landau levels or subbands. This makes it possible to study various resonance phenomena within a band.

An example of the phenomena that can be studied is the Raman scattering of light by free carriers localized at Landau levels. The principal parameters of the energy-band structure can be deduced very accurately from this scattering.

The influence of pressure on a semiconductor is manifested primarily by a change in the forbidden-band width. However, high pressures also cause other important changes in the energy structure because they affect not only the absolute extrema, which govern the forbidden band width, but also the secondary extrema. Therefore, the absolute minimum of the conduction band may shift under pressure. This makes it possible to study higher minima. Moreover, high pressures may alter significantly the density of states in an allowed band.

*Thesis submitted for the degree of Candidate of Physicomathematical Sciences, defended in 1972 at the P. N. Lebedev Physics Institute, Academy of Sciences of the USSR, Moscow.

1

TABLE 1. Spectral Range of Radiation Emitted by Semiconductor Lasers

No.	Material	λ, μ	$\hbar\omega$, eV	Excitation method
1	ZnS	0.33	3.8	O E
2	ZnO	0.37	3.4	E
3	$Zn_{1-x}Cd_xS$	0.32—0.49	3.82—2.5	O
4	ZnSe	0.46	2.7	E
5	CdS	0.49	2.5	O E
6	ZnTe	0.53	2.3	E
7	GaSe	0.59	2.1	E
8	$CdSe_{1-x}S_x$	0.49—0.68	2.5—1.8	O E
9	CdSe	0.675	1.8	O E
10	$Al_{1-x}Ga_xAs$	0.63—0.90	2.0—1.4	I
11	$GaAs_{1-x}P_x$	0.61—0.90	2.0—1.4	E I
12	CdTe	0.785	1.6	E
13	GaAs	0.83—0.91	1.50—1.36	O E I A
14	InP	0.91	1.36	I A
15	$GaAs_{1-x}Sb_x$	0.9—1.5	1.4—0.83	I
16	$CdSnP_2$	1.01	1.25	E
17	$InAs_{1-x}P_x$	0.9—3.2	1.4—0.39	I
18	GaSb	1.55	0.80	E I
19	$In_{1-x}Ga_xAs$	0.85—3.1	1.45—0.40	I
20	Cd_3P_2	2.1	0.58	O
21	InAs	3.1	0.39	O E I
22	$InAs_{1-x}Sb_x$	3.1—5.4	0.39—0.23	I
23	$Cd_{1-x}Hg_xTe$	3—15	0.41—0.08	O E
24	Te	3.72	0.334	E
25	PbS	4.3	0.29	O E I
26	InSb	5.2	0.236	O E I A
27	PbTe	6.5	0.19	O E I
28	$PbS_{1-x}Se_x$	3.9—8.5	0.32—0.146	O E I
29	PbSe	8.5	0.146	E I
30	$Pb_{1-x}Sn_xTe$	6—28	0.209—0 045	O E I
31	$Pb_{1-x}Sn_xSe$	8—31.2	0.155—0,040	I

Note. The notation used in the fifth column is as follows: O is optical pumping, E is electron-beam pumping, I is carrier injection, A is avalanche breakdown.

In view of these radical changes in the energy structure of a crystal under the influence of a magnetic field or pressure, we can naturally expect a strong dependence of the characteristics of the radiation emitted by a crystal on the applied magnetic field or pressure.

With these points in mind it would be interesting to study the dependence of the emission frequency of semiconductor lasers on the applied pressure and magnetic field because this would give information on the energy spectrum and also help in designing high-power tunable sources of monochromatic radiation. We shall now consider in greater detail the problem of tunability of coherent radiation frequency.

Currently available lasers can emit radiation of wavelengths ranging from near ultraviolet to far infrared. The active media used in lasers are solids, including semiconductors [1-7], liquids, and gases.

Table 1 gives the published information (see, for example, [8]) on the characteristics of currently available lasers. An examination of this table shows that semiconductor lasers emit over a wide spectral range. For example, zinc sulfide excited by electron bombardment emits coherent radiation at $\lambda = 0.33$ μ [9], whereas the $Pb_{1-x}Sn_xSe$ (x = 0.19) injection laser emits radiation with $\lambda \approx 31$ μ and can be tuned by a magnetic field up to 34 μ [10].

There are several ways of ensuring tunability of the coherent radiation frequency.

The first method is to increase the range of available frequencies by a suitable selection of new active media and by utilization of all possible transitions in the emission spectra of these media. The number of semiconducting materials which can emit coherent radiation is about thirty. However, this number is found to be much larger if we include also lasers made of solid solutions of some semiconductor compounds (ZnCdS, CdSeS, AlGaAs, GaAsP, GaAsSb, InAsP,

InGaAs, CdHgTe, PbSSe, PbSnSe, PbSnTe). Usually such solutions form a continuous series. Therefore, we can vary monotonically the forbidden band width by varying the solid-solution composition and thus obtain a wide range of emission frequencies.

The second method is to use nonlinear interactions between radiation and matter (generation of harmonics, sum, and difference frequencies; stimulated Raman scattering, and so on).

The two methods described above provide only a discrete set of emission frequencies and, therefore, they do not solve the problem of continuous tuning of the coherent radiation frequency.

The third method allows us to tune the emission frequency continuously and it utilizes parametric interactions involving light waves in optically transparent nonlinear media. This method makes it possible to vary the stimulated emission frequency within a fairly wide range, but the low conversion efficiency and stringent requirements in relation to the optical quality and stability of nonlinear elements are serious disadvantages.

The fourth method for the tuning of the laser emission frequency is based on the use of tunable dispersive (selective) resonators, which make it possible to vary the laser radiation frequency continuously and without significant losses within the limits of the emission spectrum of the active medium. However, it is difficult to use this method in the case of conventional semiconductor lasers (with Fabry-Perot resonators formed by the faces of the crystal). Nevertheless, mode selection and wavelength tuning within a narrow range have been achieved [11] for a GaAs laser with a composite resonator.

Finally, the fifth method of frequency tuning is based on the modification of the energy levels of an active medium by varying the temperature of this medium or by applying pressure, or electric [12-14] and magnetic fields. This "direct" shift of the levels by external agencies is a particularly promising method in the case of semiconductor lasers because the energy levels of semiconductors can be varied to the greatest extent.

A considerable frequency tuning by magnetic fields [15-17], hydrostatic pressure [18, 19], and temperature [20-22] has been demonstrated for semiconductor injection lasers. The application of strong magnetic fields to semiconductors has made it possible to tune the stimulated Raman scattering accompanied by electron-spin flip [23-25]. The latter effect provides a very promising method for continuous frequency tuning.

External agencies can also be used to vary quite strongly the refractive index of semiconductors, which provides a means for continuous displacement of the emission frequencies of individual modes [19, 26-29].

It should be noted that semiconductor lasers have certain advantages over lasers utilizing glasses, ionic crystals, and gases and these advantages include small size, high efficiency, opportunity for direct conversion of electrical energy into coherent radiation, and consequent simplicity of the control of the radiation intensity. However, the output power of semiconductor lasers is considerably lower than that of lasers of other types.

It follows from the above review that the most effective methods for the tuning of the emission frequency of semiconductor lasers are hydrostatic pressures and magnetic fields. The strong temperature dependences of the threshold conditions limit somewhat the use of temperature for tuning purposes.

The purpose of the present investigation was to determine the influence of magnetic fields and hydrostatic pressures on the radiation emitted by II-V and IV-VI semiconductor lasers and to examine the possibilities of building tunable infrared sources of monochromatic radiation.

Investigations in magnetic fields were carried out on semiconductors with small carrier masses (InAs, InSb, and PbSe), and in these cases the widest range of frequency tuning was obtained. The influence of pressures was studied in the case of lasers operating at liquid nitrogen temperature and characterized by large positive (GaAs) and negative (PbSe) pressure coefficients of the forbidden-band width.

Chapter I is concerned with the basic physical ideas on the band quantization in magnetic fields, influence of pressure on the forbidden-band width, and influence of external agencies on the emission frequencies of semiconductor lasers.

Chapter II describes the method developed for optical investigations in strong magnetic fields and under high hydrostatic pressures. A description is given of a Q-switched CO_2 laser used as an excitation source in a study of the stimulated Raman scattering of light in indium antimonide. A construction of an optical cryostat with a superconducting solenoid for the use in the infrared part of the spectrum is described. A report is made of the method and apparatus used for optical investigations in the infrared range under hydrostatic pressures at 77°K. Infrared radiation detectors and a technique for scanning the radiation emitted from samples are described.

Chapter III gives the results of an investigation of the spontaneous and coherent emission from InAs injection lasers in weak and strong magnetic fields, of the dependence of the emission frequency of InSb lasers on magnetic fields, and of the spontaneous and coherent emission from PbSe diodes in magnetic fields. The effective carrier masses and the g factors are deduced from the experimental results. The emission regions of double-injection InSb lasers are given. The ranges of tuning in magnetic fields up to 56 kOe are determined. The dispersion of the refractive index in magnetic fields is deduced from the dependence of the emission frequency of individual modes emitted by InAs, InSb, and PbSe lasers.

Chapter IV is devoted to the magnetically tunable stimulated Raman scattering of light by free electrons in InSb. This effect was observed for the first time in the USSR (after the work reported in [3, 24]). The inelastic scattering of light involves the excitation of an electron from a lower to an upper Landau spin sublevel. The dependence of the scattered-like frequency on the magnetic field is used to find the g factor of electrons in InSb. A strong dependence of the g factor on the magnetic field and electron density is attributed to the band nonparabolicity. The scattered light frequency is found to be tunable in fields from 30 to 56 kOe.

Chapter V gives the results of an investigation of the influence of pressure on the emission from lead selenide and gallium arsenide semiconductor lasers. Application of hydrostatic pressures up to 5 kbar alters the emission wavelength of the PbSe laser from 7.34 to 9.63 μ. Moreover, a continuous shift of the frequencies of the individual modes is found in a narrow range of pressures (~300 bar). In the case of the GaAs laser the application of pressures up 8.8 kbar alters the emission frequency in the range $\lambda = 0.84-0.79$ μ. The pressure coefficients of the forbidden-band widths of GaAs and PbSe are determined.

The concluding chapter is concerned with possible future developments of magnetooptic investigations and experiments under high hydrostatic pressures in the infrared part of the spectrum. Possible applications of semiconductor lasers tunable by pressure or magnetic field are indicated.

INFLUENCE OF MAGNETIC FIELDS AND HIGH PRESSURES ON ENERGY SPECTRA OF SEMICONDUCTORS

§ 1. Influence of Magnetic Fields on Energy

Structure of III-V and IV-VI Semiconductor

Compounds

A. Theoretical Assumptions. A charged particle subjected to a static homogeneous magnetic field H moves down the field along a helical path of radius

$$r = v_\perp / \omega_c, \tag{1}$$

where $v_\perp$ is the component of the velocity in a plane perpendicular to the magnetic field, and

$$\omega_c = eH / m_0 c \tag{2}$$

is the angular rotation frequency of the particle in that plane, known as the cyclotron frequency; e and m_0 are, respectively, the charge and mass of the particle; c is the velocity of light.

The quantum effects are important at low temperatures and in strong magnetic fields. In this case the solution of the Schrödinger equation for a system of electrons in a crystal predicts quantization of allowed bands into Landau levels.

In the case of semiconductors whose dispersion law $\mathscr{E}(\mathbf{k})$ is quadratic and whose constant-energy surfaces are ellipsoids of revolution, a three-dimensional allowed band splits into a series of one-dimensional Landau bands, each of which breaks up into two subbands corresponding to the opposite directions of spin. Then, the energy of a carrier in a band is of the form [30]

$$\mathscr{E} = \hbar^2 k_H^2 / 2 m_H + \hbar \omega_c (l + 1/2) + s g \beta H. \tag{3}$$

Here, $\hbar k_H$ and m_H are, respectively, the quasimomentum and the effective mass in the direction of the magnetic field; $\hbar$ is the Planck constant; l is the oscillatory quantum number which assumes integral values (l = 0, 1, 2, ...); s is the quantum number associated with the electron momentum and equal to ±1/2; $\beta = e\hbar / 2 m_0 c$ is the Bohr magneton; g is the spectroscopic splitting factor, which is given by the following expression in the case of ellipsoidal constant-energy surfaces:

$$g = \pm (g_\parallel^2 \cos^2 \varphi + g_\perp^2 \sin^2 \varphi)^{1/2}, \tag{4}$$

where $g_\parallel$ and $g_\perp$ represent the longitudinal (along the ellipsoid axis) and transverse values of the g factor; φ is the angle between the axis of revolution of the ellipsoid and the applied magnetic field. It should be noted that the expression (3) for the cyclotron frequency ω_c does not contain the free-electron mass m_0 but the cyclotron effective mass m_c, which is given by

$$m_c = m_\perp \left(\frac{K}{K \cos^2 \varphi + \sin^2 \varphi} \right)^{1/2}, \tag{5}$$

where $K = m_\parallel / m_\perp$ is the anisotropy coefficient; $m_\parallel$ and $m_\perp$ are the longitudinal and transverse masses.

The trajectory of an electron in the **k** space can be found because it represents the line of intersection of a constant-energy surface by a plane perpendicular to the magnetic field. The expression (3) shows that the periodic motion in this plane is quantized and is described by wave functions of a harmonic oscillator, whereas the motion along the magnetic field is described by a plane wave.

A strong field means that an electron or a hole in a crystal makes only one or several revolutions about the magnetic field axis before it is scattered during a relaxation time τ, i.e.,

$$\omega_c \tau = \frac{eH}{m_c c}\tau = \frac{\mu H}{c} > 1, \tag{6}$$

where μ is the carrier mobility. Moreover, the thermal broadening of the Landau levels should be slight, i.e., the cyclotron energy should exceed the thermal energy of carriers:

$$\hbar\omega_c > k_0 T, \tag{7}$$

where T is the temperature and k_0 is the Boltzmann constant.

When the spin splitting, which is usually about half the cyclotron splitting, is allowed for, the criterion (7) becomes more stringent and in cases of practical importance it assumes the form

$$g\beta H \geqslant 2k_0 T. \tag{7a}$$

Thus, the criterion of strong magnetic fields is satisfied most easily at low temperatures in the case of semiconductors with small effective masses and high carrier mobilities.

The quantization of energy levels in magnetic fields alters considerably the density of states. This density $\rho(\mathcal{E})$ can be found using Eq. (3) and it is described by [30]

$$\rho(\mathcal{E}) = \frac{(2m_H)^{1/2}m_c\omega_c}{(2\pi\hbar)^2}\sum_{l,s}\left[\mathcal{E} - \hbar\omega_c\left(l + \frac{1}{2}\right) - sg\beta H\right]^{-1/2}. \tag{8}$$

We can see that in a magnetic field the density of states is a very nonmonotonic function of the energy. When this energy is $\mathcal{E} = \hbar\omega_c\left(l + \frac{1}{2}\right) + sg\beta H$, the density of states becomes infinite. In the intervals between these infinite singularities, the density $\rho(\mathcal{E})$ decreases but remains proportional to the magnetic field. This behavior is illustrated in Fig. 1.

The recombination radiation emitted from semiconductor lasers is due to transitions between bands with degenerate carrier distributions or at least between a degenerate band

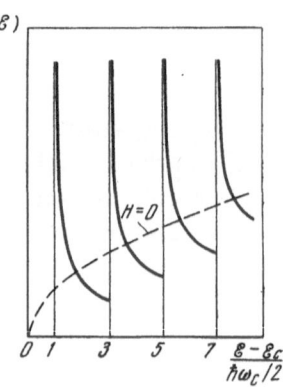

Fig. 1. Density of energy states in a quantizing magnetic field. The dashed curve shows the density in the absence of the field.

and an impurity. Variation of the magnetic field causes the Landau bands to move relative
to the Fermi level and the density of states near this level oscillates. Since coherent recombination radiation is due to transitions close to the Fermi level, these density-of-states
oscillations give rise to oscillations in the emission (absorption) spectra. If radiative transitions take place between the valence and conduction bands whose edges are at the same point
in the **k** space and are only spin-degenerate, the energy of a photon emitted in a magnetic field
and corresponding to maxima in the recombination radiation spectrum is [see Eq. (3) for $\mathbf{k}_H = 0$]

$$\hbar\omega = E_g + \hbar\omega_{cn}\left(l_n + \frac{1}{2}\right) + \hbar\omega_{cp}\left(l_p + \frac{1}{2}\right) + s_n g_{ni}\beta H + s_p g_p \beta H, \tag{9}$$

where E_g is the forbidden-band width. The quantum numbers should then satisfy the following
selection rules: $\Delta l = l_n - l_p = 0$ and $\Delta s = s_n - s_p = 0, \pm 1$. In the case of $\Delta s = 0$ transitions at
right angles to H, the emitted light is linearly polarized and the oscillations of the electric
vector are directed along H(π component). In the case of the $\Delta s = \pm 1$ transitions, the emitted
radiation is circularly polarized in a plane perpendicular to the magnetic field, and the two
signs in this selection rule correspond to two different directions of the circular polarization,
represented by the σ_r and σ_l components. If the recombination radiation is observed at right-
angles to the magnetic field, the superposition of the σ_r and σ_l spectra gives rise to a σ spectrum, which is polarized linearly at right angles to the magnetic field.

Since in a quantizing magnetic field the density of states is proportional to H, an increase in the magnetic field intensity lowers the Fermi level. In the quantum limit, when the
Fermi level lies between the two lowest magnetic levels, this may give rise to a lifting of the
degeneracy at specific values of H.

If electrons fill only the lowest spin Landau sublevel, the dependence of the Fermi level
on the magnetic field is given by [31]

$$F(H) = \frac{\hbar\omega_c}{2}\left[1 - \frac{|g|}{2}\frac{m}{m_0} + \frac{4}{9}\left(\frac{2F_0}{\hbar\omega_c}\right)^3\right], \tag{10}$$

where F_0 is the Fermi energy in the absence of the field. Since the bottom of the conduction
band rises in a magnetic field by an amount $\frac{\hbar\omega_c}{2}\left(1 - \frac{|g|}{2}\frac{m}{m_0}\right)$, the criterion of the electron gas
degeneracy is of the form

$$\frac{4}{9}\frac{F_0}{k_0 T}\left(\frac{2F_0}{\hbar\omega_c}\right)^2 \gg 1 \tag{11}$$

and it depends on the magnetic field.

If electrons fill both spin sublevels of the lowest Landau level, the Fermi level can be
found from the relationship [31]

$$F(H) = \frac{\hbar\omega_c}{2}\left[1 + \frac{1}{9}\left(\frac{2F_0}{\hbar\omega_c}\right)^3 + \frac{9}{16}\left(\frac{\hbar\omega_c}{2F_0}\right)^3\left(\frac{gm}{m_0}\right)^2\right]. \tag{10a}$$

In this case the degeneracy criterion is

$$\frac{9\,(\hbar\omega_c)^4}{256 k_0 T F_0^3}\left[\frac{4}{9}\left(\frac{2F_0}{\hbar\omega_c}\right)^3 + |g|\frac{m}{m_0}\right]^2 \gg 1. \tag{11a}$$

It should be pointed out that Eqs. (10) and (10a) are valid for an infinite density of states
at each of the Landau levels. Allowance for the finite-level width weakens the degeneracy
criteria (11) and (11a).

Carriers in many semiconductors do not obey the quadratic dispersion law. This is particularly true of the narrow-gap semiconductors such as InSb, PbSe, and others, which are characterized by a strong interaction between the conduction and valence bands. As shown by Kane [32], the dependence of the energy E on the quasimomentum p in InSb can be expressed in the form

$$\mathcal{E} = \sqrt{\frac{p^2}{2m} E_g + \left(\frac{E_g}{2}\right)^2} \approx \frac{E_g}{2} + \frac{\hbar^2 k^2}{2m} - \frac{\hbar^4 k^4}{4m^2 E_g}, \tag{12}$$

where the energy is measured from the middle of the forbidden band and m represents the effective mass at the band edge. Many experiments have shown that the Kane law applies also to other III-V semiconductors. A generalized Kane model developed for the case of ellipsoidal constant-energy surfaces is applicable also to some lead chalcogenides (PbSe, PbS) [30].

Another consequence of the strong interaction between the conduction and valence bands is the large magnetic moment of electrons, which is equivalent to a large value of the spectroscopic splitting factor g. The Kane theory for InSb [33] gives the formula

$$g = 2\left[1 - \left(\frac{m_0}{m} - 1\right)\left(\frac{\Delta}{3E_g + 2\Delta}\right)\right], \tag{13}$$

where Δ is the spin-orbit splitting; m is the effective mass at the band edge. Using the Kane dispersion law (12), we can also apply Eq. (13) to electron wave vectors $k \neq 0$ [34].

In the case of PbSe-type semiconductors, for which the extrema of the conduction and valence bands are only spin-degenerate (see subsection B), the two-band Kane model [35, 36] gives

$$g = 2\frac{m_0}{m_c}, \tag{14}$$

where the cyclotron effective mass m_c is given by Eq. (5).

B. Energy Structure of Compounds GaAs, InAs, InSb, and PbSe. Before considering the influence of magnetic fields and high pressures on the energy spectra of semiconductors, it is useful to discuss their band structure first.

Compounds of the III-V type crystallize in cubic lattices of the zinc blende type. The atomic bonds in the lattice are covalent with some admixture of ionicity.

The energy band structure of gallium arsenide is shown in Fig. 2 [37]. Indium arsenide and antimonide have similar band structures. The smallest energy gap between the valence and conduction bands is located at the center of the Brillouin zone (point Γ). The valence band is degenerate at this point. It consists of a light-hole band, a heavy-hole band, and a band

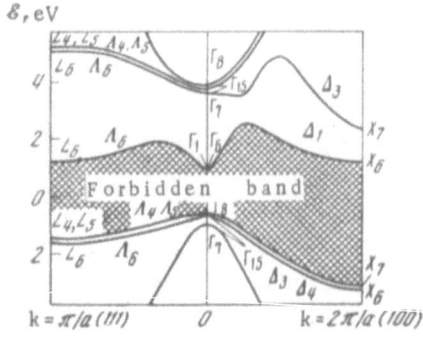

Fig. 2. Energy band structure of GaAs [37].

TABLE 2. Some Properties of GaAs, InAs, InSb, and PbSe

Property	GaAs	InAs	InSb	PbSe
Lattice constant, Å	5.65	6.06	6.48	6.12
Effective charge, e	0.46	0.49	0.46	0.59
E_g, eV (at 77°K)	1.510	0.41	0.228	0.176
E_g^0, eV (at 4°K)	1.517	0.425	0.236	0.165
$(\partial E_g / \partial T)_p \times 10^4$, eV/deg	—5.0	—3.5	—2.8	+4.5
Refractive index n	3.59	3.69	3.96	5.3
$(1/n)(dn/dT) \times 10^4$, deg^{-1}	0.45		0.27	3.0
Bulk compressibility $\varkappa \times 10^6$, bar^{-1}	—1.25	—1.72	—2.14	—2.07
Linear thermal expansion coefficient $\alpha_0 \times 10^6$ at 77°K, deg^{-1}	+3.64	+3.40	—0.06*	+16
Energy of transverse optical phonons, meV	33.5	27.5	23.5	5.5
Energy of longitudinal optical phonons, meV	35.5	30.6	25.5	16.5
High-frequency permittivity ε_∞	10.90	11.8	15.68	25.2
Static permittivity ε_0	12.53	14.55	17.72	227

*At T = 10°K.

split off by the spin—orbit interaction (the splitting is 0.33 eV). However, the valence-band maxima are not located exactly at the center of the Brillouin zone but are shifted away from it by a short distance. The conduction band is spin-degenerate and it has also two additional minima along the [100] and [111] directions. At the point Γ the constant-energy surfaces of the valence and conduction bands are spherical. The strong interaction between these bands results in a deviation of the bottom of the conduction band from the parabolic law. This effect is particularly strong in the case of semiconductors with narrow forbidden bands (InSb, InAs).

Some properties of GaAs, InAs, and InSb are given in Tables 2 and 3 [38, 39].

Lead selenide is a semiconductor compound with a considerable proportion of ionicity. It crystallizes in an fcc lattice of the NaCl type

The results of many experiments have shown that the constant-energy surfaces of the conduction and valence bands of lead selenide are eight ellipsoids elongated along the ⟨111⟩ directions and the centers of these ellipsoids are located at the points L in the Brillouin zone. The energy band structure of PbSe calculated by Herman et al. [40] is shown in Fig. 3. The minimum energy gap between the valence and conduction bands is located at the point L in the Brillouin zone, i.e., direct optical transitions play an important role in the recombination of carriers in lead selenide. The conduction and valence bands are only spin-degenerate. There are two additional conduction bands and two valence bands at the point L and they are separated from the main bands by gaps of 1.5-2 eV.

In calculating the energy band structure of lead chalcogenides [40] it is necessary to allow not only for the spin—orbit interaction but also for other relativistic corrections to the

TABLE 3. Effective Masses and g Factors at Band Edges and Spin — Orbit Splittings of GaAs, InAs, and InSb

Property	GaAs	InAs	InSb
m_n/m_0	0.071	0.024	0.013
m_{p1}/m_0	0.5	0.41	~0.6
m_{p2}/m_0	0.12	0.025	0.015
g_n	0.523	—15	—50.7
Δ, eV	0.33	0.43	0.98

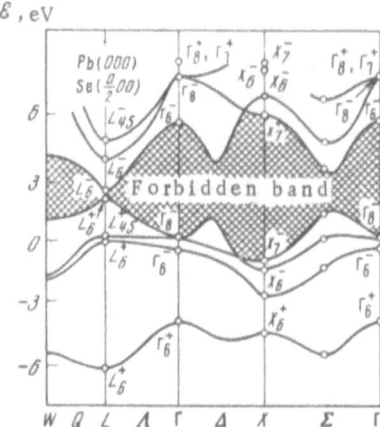

Fig. 3. Energy band structure of PbSe [40].

Hamiltonian. The strong spin–orbit interaction and the large relativistic corrections in the case of lead chalcogenides are due to the high atomic number of lead atoms. Moreover, observations of the Knight shift indicate that at the edge of the valence band of PbTe an electron located near a lead atom is described by a wave function which is of the same nature as the atomic s function, and this should be taken into account in calculations.

The spin–orbit mixing of the two L_6^+ states of the valence band and the two L_6^- states of the conduction band is responsible for the finite matrix element of the transition between L_6^+ and L_6^- for all directions of the wave vector **k** of an electron near the point L. Experiments have shown that the radiative recombination in lead chalcogenides is the dominant mechanism at low temperatures.

The properties of lead selenide are given in Tables 2 and 4 [41–43]. It is worth noting the high static permittivity ε_0 of PbSe, which is due to the small energy of transverse optical phonons. The high value of ε_0 and the small effective mass are responsible for the merging of impurity levels in PbSe with an allowed band so that one common band is formed. It follows that impurities or vacancies are ionized even at hellium temperatues and the carrier density is independent of temperature.

It is clear from Tables 3 and 4 and from Eqs. (3)–(5) that the greatest shift of the cyclotron and spin levels in strong magnetic fields may be exprected for electrons in InSb and InAs and also for electrons and holes in PbSe. However, since the energy bands of these compounds are nonparabolic, in the case of degenerate crystals the effective mass of a carrier at the Fermi level is larger and the g factor is smaller than at the band edge. Therefore, the splitting is somewhat smaller than expected.

We shall now give some estimates applicable to typical conditions in the investigated InAs, InSb, and PbSe lasers.

It follows from Eq. (6) that a magnetic field is quantizing in InAs for $H \gtrsim 10$ kOe and $\mu_n \approx 10^4$ cm$^2 \cdot$ V$^{-1} \cdot$ sec^{-1}, in InSb for $H \gtrsim 1$ kOe and $\mu_n \approx 10^5$ cm$^2 \cdot$ V$^{-1} \cdot$ sec^{-1}, and in PbSe for

TABLE 4. Effective Masses and g Factors at Band Edges of PbSe [42, 43]

Band	$m_\perp$	$m_{||}$	K	$|g_{||}|$	$g_{[100]}$
Conduction band	0.040±0.008	0.07±0.015	1.75±0.2	27±7	∓22±4
Valence band	0.034±0.007	0.068±0.015	2.0±0.2	32±7	∓22±4

H ⩾ 1 kOe and $\mu_{n, p} \approx 10^5$ cm$^2 \cdot$ V$^{-1} \cdot$ sec^{-1}. The mobilities just quoted are taken at helium temperatures.

According to Eq. (7a), the spin splitting in magnetic fields may be observed for InAs in fields H ⩾ 20 kOe at T ≈ 10°K, for InSb in H ⩾ 6 kOe at T ≈ 20°K, and for PbSe in H ⩾ 3 kOe at T ≈ 4.2°K.

In the case of indium antimonide with a bulk electron density (initial or established by double injection) n = 1.6×10^{16} cm^{-3} the effective mass is $m_n = 0.016 m_0$ [39] and, consequently, in the absence of a magnetic field the Fermi energy is F = 14 meV. We shall use $\mathcal{E}_1$ and $\mathcal{E}_2$ to denote the lower and upper Landau spin sublevels (l = 0), measured from the bottom of the conduction band $\mathcal{E}_c$ in H = 0. Then, it follows from Eq. (3) that in a field H = 20 kOe, we have $\mathcal{E}_1$ = 4.4 meV and $\mathcal{E}_2$ = 9.7 meV, i.e., $\mathcal{E}_1 < \mathcal{E}_2 < F_0$. According to Eq. (10a), the Fermi level in a magnetic field H = 20 kOe lies at 13.5 meV, again measured from $\mathcal{E}_c$ in H = 0. If the Fermi level is measured from the bottom of the conduction band in H = 20 kOe, it is found that F'(20 kOe) − $\mathcal{E}_1$ = 9.1 meV. Thus, the electron gas is degenerate up to T ≈ 50°K.

If the magnetic field is increased to H = 50 kOe, we find that $\mathcal{E}_1$ = 10.9 meV and $\mathcal{E}_2$ = 23.1 meV, i.e., $\mathcal{E}_1 < F_0 < \mathcal{E}_2$. According to Eq. (10), the Fermi level at 50 kOe is F (50 kOe) = 15.1 meV and F'(50 kOe) = 4.2 meV. Consequently, once again the gas is degenerate at T ≤ 25°K.

Similar estimates for InSb with an electron density n = 5.6×10^{15} cm^{-3} show that in a field H = 20 kOe, we have $\mathcal{E}_1 < F_0 < \mathcal{E}_2$ and F'(20 kOe) = 3.2 meV, i.e., we can expect the degeneracy to be lifted at T ⩾ 20°K. However, in a field H = 50 kOe, we have F'(50 kOe) = 0.7 meV and the degeneracy should be lifted at T ⩾ 8°K.

As mentioned earlier, allowance for the finite width of the Landau levels results in lifting of the degeneracy at higher temperatures or in stronger magnetic fields.

§2. Influence of Pressure on Energy Structures of

III-V and IV-VI Compounds

The application of a pressure to a semiconductor alters the lattice constants so that the band structure is affected. A simple assumption that the overlap of the wave functions of electrons increases with increasing pressure because of a reduction in the atomic distances, which leads to an expansion of the allowed bands and narrowing of the forbidden band, is not always justified. In fact, experiments show that the forbidden band of a semiconductor may decrease or increase with pressure. The governing factor is the energy band structure of a semiconductor along different crystallographic directions.

In the case of elemental group IV semiconductors and in III-V compounds the valence-band maximum is located at the point Γ in the k space (Fig. 2). This maximum has a complex structure which changes considerably when a cubic crystal is subjected to uniaxial deformation. A hydrostatic pressure does not disturb the cubic symmetry and it has very little effect on the valence band. Therefore, changes in the energy band structure of a semiconductor under the influence of hydrostatic pressures are usually considered only at the top of the valence band.

It has been established theoretically and experimentally that the semiconductors under consideration have three types of the lowest minima in the conduction band: one type is located at the center of the Brillouin zone at the point Γ (**k** = 0); the other type is oriented along the [100] axis and it lies close to the point X in the Brillouin zone (**k** = $2\pi/a$); the third type is oriented along the [111] axis and lies at the point L (**k** = π/a) (Fig. 2). The coefficients representing the changes in the position of these minima with increasing pressure (pressure coefficients) differ in sign and magnitude and, according to the empirical rule of Paul [44], these coefficients are approximately +12, +5, and −1.5 × 10^{-6} eV/bar at the points Γ, L, and X, respectively.

The experimentally determined change in the forbidden-band width is generally a linear function of pressure, in accordance with the linear theory of the deformation potential [45]. However, in some cases two linear regions with different slopes meet at a kink. Such a kink is exhibited, for example, by germanium and it is evidence of a change in the type of minimum which governs the forbidden-band width. These effects can occur in group IV semiconductors and in III-V compounds because their conduction-band minima are frequently separated by relatively small energy intervals.

The lowest minima of the conduction band of lead selenide are localized at the point L along the [100] direction (points Δ) and along the [110] axis (points Σ) (Fig. 3). The energy gaps between the minima are too large (~ 1 eV) for the observation of the change in the type of the absolute minimum.

Table 5 gives the pressure coefficients of the best-known semiconductors [19, 39, 46]. It is clear from this table that in the case of materials with the same crystallographic symmetry the pressure coefficients for each type of minimum have the same sign and approximately the same magnitude. It is worth noting that in the case of lead chalcogenides the pressure coefficient at the point L is negative. On the other hand, the temperature coefficient of the forbidden band width of PbSe is positive (Table 2). This has recently been explained by the band inversion [47].

Table 5 gives the data necessary for the selection of the materials to be investigated. It is known that the laser effect occurs if the minimum of the conduction band is located at the same point in the **k** space as the maximum of the valence band. This "direct" band structure is typical of III-V compounds and of lead chalcogenides. Moreover, the highest positive pressure coefficient at the point Γ is exhibited by III-V compounds, whereas lead chalcogenides exhibit the highest negative pressure coefficient at the point L. Thus, lasers prepared from these materials should have the widest frequency tuning range under pressure. These considerations and allowance for the working temperature of lasers were the reasons why lead selenide and gallium arsenide were selected for our investigation.

We must distinguish two separate effects due to pressure: one is the change of the energy bands relative to one another and the other is the distortion of the band edges. The first effect alters the properties which are associated with the interband transitions, such as the position

TABLE 5. Pressure Coefficients of Forbidden Band Width for Three Types of Minima in Conduction Band at Γ, X, and L

Material	$(\partial E_g/\partial P)_T \cdot 10^{-6}$ eV/bar		
	Γ	X	L
C	—	1.0	—
Si	—	—1.5	5.0
Ge	12.0	—2.0	5.0
Sn	—	—	5.0
AlSb	—	—1.6	—
GaP	10.7	—1.1; —1.7; —1.8	—
GaAs	9.4; 12.0; 11.7	—1	—
InP	4.6; 8.4	—10.0 (?)	—
GaSb	12.0; 16.0; 15.0	—1.5	5.0; 7.3: 4.4
InAs	4.8; 8.5; 5.5; 10.0	—	3.2
InSb	14.2; 15.5	—	5
PbS	—	—	—9.15
PbSe	—	—	—9.1
PbTe	—	—	—7.4
CdTe	8	—	—
ZnTe	6		

Note. Different values of pressure coefficients for the same material are taken from different sources.

of the fundamental absorption edge or the recombination radiation maximum, and it also affects the distribution of electrons between the allowed bands and impurity levels (this alters the position of the Fermi level and the intrinsic carrier density).

The second effect changes the density of states in a band, i.e., it alters the effective carrier mass. If this mass in the bands is governed only by the interaction between the conduction and valence bands, the mass near an extremum varies linearly with the forbidden-band width so that we have

$$\left(\frac{\partial \ln E_g}{\partial P}\right)_T = \left(\frac{\partial \ln m}{\partial P}\right)_T. \tag{15}$$

In fact, the relative change in the effective masses with pressure differs somewhat from the relative change in the forbidden-band width. This may be due to the contribution of the interaction with other nearby bands to the effective mass. Secondly, in the case of samples with relatively high carrier densities the measured effective mass corresponds not to an extremum but to the Fermi level, and at this level the effective mass depends less strongly on the forbidden-band width.

It is known [48] that the temperature dependence of the forbidden-band width is associated with the thermal expansion of a crystal and with the interaction between carriers and lattice vibrations. The derivative $\partial E_g/\partial T$ can be represented in the form

$$dE_g/dT = (\partial E_g/\partial T)_0 + (\partial E_g/\partial T)_V, \tag{16}$$

where $(\partial E_g/\partial T)_0$ is the temperature-induced change of the forbidden-band width in a rigid lattice (it arises because of thermal expansion), whereas $(\partial E_g/\partial T)$ is the temperature-induced change in E_g for a crystal with a fixed volume (this contribution is due to the interaction with phonons).

The first term in Eq. (16) can be expressed in terms of the pressure coefficient of E_g by means of the thermodynamic relationship

$$(\partial E_g/\partial T)_0 = -(3\alpha_0/\varkappa)(\partial E_g/\partial P)_T, \tag{17}$$

where $\varkappa$ and α_0 are the compressibility and the linear thermal expansion coefficient.

It is clear from the above expression that measurements of the isothermal pressure coefficient of the forbidden-band width should make it possible to estimate the relative influence of the thermal expansion and of the interaction with phonons on dE_g/dT. Substituting in Eq. (17) the necessary values for GaAs and PbSe (Table 2 and 5), we find that $(\partial E_g/\partial T)_0 \approx 1 \times 10^{-4}$ and 2.1×10^{-4} eV/deg, respectively. Comparing these values with the values of dE_g/dT found from optical measurements (Table 2), we can see that the thermal expansion of the lattice accounts only for one-fifth of the temperature coefficient of the forbidden-band width of GaAs and about one-half in the case of PbSe.

§3. Characteristics of Semiconductor Laser Operation

Affected by Variation of Temperature, Pressure, and

Magnetic Field

A semiconductor laser is a crystal in which nonequilibrium electrons and holes are generated by external excitation (electric-current injection, electron-beam bombardment, or

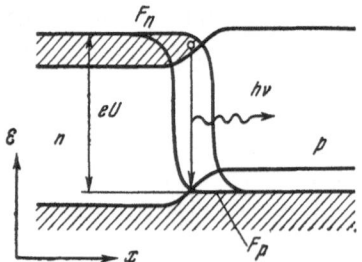

Fig. 4. Doped (to degeneracy) p−n junction, subjected to a forward bias U.

illumination) [2, 7, 49-51]. In order to ensure a high quantum efficiency of recombination radiation, we must use semiconductors with the direct band structure* and the laser action can
be observed if the following two conditions are satisfied.

The first condition is the population inversion [2]:

$$F_n + F_p > \hbar\omega \approx E_g, \tag{18}$$

where F_n, and F_p are the quasi-Fermi levels of electrons and holes, respectively, and ω is the
output radiation frequency.

The second condition demands that the recombination radiation should exceed all the
active losses in the semiconductor, i.e.,

$$R \exp[(g_s - \alpha) \cdot l] > 1, \tag{19}$$

where R is the reflection coefficient, at the frequency ω, of the surface separating the semiconductor from the ambient medium; g_s and α are the gain and absorption coefficient of the
active medium; l is the separation between the reflecting faces of the laser crystal.

In the case of lasers with p−n junctions, the conditions (18) requires the presence of an
impurity in a concentration sufficient to ensure that the density of electrons in the n-type region and of holes in the p-type region both lie in the electron-gas degeneracy range ($\sim 10^{-18}$
cm^{-3}). In this situation the Fermi level lies, respectively, in the conduction or valence band,
and a population inversion is achieved by biasing the p−n junction in the forward direction
(Fig. 4). When the applied voltage U reaches the contact potential U_c ($U_c \approx E_g$), holes are
injected across the p−n junction into the n-type region (and, conversely, electrons are injected
into the p-type region). The injected carriers recombine producing radiation of energy $\hbar\omega$
and stimulated emission is observed when the injection level exceeds the threshold current
density j_{th}. Generation of radiation in an injection laser is localized in a narrow region where
the absorption is negative. This region is located at the p−n junction plane and it is surrounded by the n- and p-type regions containing free electrons and holes. The thickness of
the active region is governed by the steepness of the diffusion front and it amounts to about
1 μ, which is considerably greater than the thickness of the space-charge region (the Debye
length for a carrier density of $\sim 10^{18}$ cm^{-3} is approximately 10^{-6} cm). The depth of penetra-
tion of light into the passive n- and p-type regions is governed by the difference between the
permittivities of the n- and p-type regions and of the p−n junction itself. The influence of
free carriers and a reduction of the absorption coefficient of light in the heavily doped p- and
n-type regions makes the permittivity of a crystal at frequencies $\omega = E_g/\hbar$ lower than in the
p−n junction. Therefore, light is concentrated in an optically denser part of the p−n junction

---

*Recently stimulated emission was observed also for GaAs$_{0.5}$P$_{0.5}$, which is a semiconductor
with the indirect band structure [52].

($\sim 10\ \mu$), which acts as a dielectric waveguide. The laser action produces transverse TE or TM modes each of which corresponds to a multitude of longitudinal modes differing from one another by the number of half-waves that can be fitted along the resonator axis.

The condition for the excitation of the longitudinal modes in a conventional Fabry–Perot resonator is

$$k \frac{\lambda}{2} = nl, \tag{20}$$

where $k = 1, 2, 3, \ldots$ is an integer; λ is the emission wavelength; n is the refractive index of the active medium which is generally a function $n = f(\lambda, x)$; l is the resonator length, which is also a function of an external agency x (x denotes temperature, pressure, or magnetic field). It follows from the condition (20) that the shift of the gain profile of the laser under the influence of an external agency results in switching of stimulated emission from a given mode to a series of other modes (under single-mode emission conditions). In the multimode emission case an external agency causes the excitation of short-wavelength modes and damping of the long-wavelength modes if λ decreases with rising x, and vice versa. This results in a discrete tuning on the emission frequency.

Continuous tuning of the emission wavelength can be achieved by shifting the energy positions of the individual modes because n and l depend on x. Bearing this point in mind, we find that Eq. (20) for a given mode gives the following expression for the relative change in the wavelength:

$$\frac{1}{\lambda} \frac{d\lambda}{dx} = \frac{\left[\frac{1}{n} \left(\frac{\partial n}{\partial x} \right)_\lambda + \frac{1}{l} \frac{dl}{dx} \right] 2nl\Delta\lambda}{\lambda^2}. \tag{21}$$

Here, $\Delta\lambda$ is the mode separation, which is found from the expression

$$\Delta\lambda = \frac{\lambda^2}{2l\,(n - \lambda dn/d\lambda)} = \frac{\lambda^2}{2\,ln^*}. \tag{22}$$

Since the relative change in the refractive index under the action of an external agency is usually much greater than the relative change in the length of a crystal, it follows that in practical cases Eq. (21) can be reduced to

$$\frac{1}{\lambda} \frac{d\lambda}{dx} \approx \frac{2l\Delta\lambda \left(\frac{\partial n}{\partial x} \right)_\lambda}{\lambda^2}. \tag{21a}$$

The dispersion of the refractive index of the active medium in an external field can be found by measuring experimentally λ and the relative shift $\Delta\lambda$ and then applying Eq. (21a).

The width of a laser emission line (corresponding to a single mode) is, in principle, infinitesimally small. However, the finite value of the Q factor of the resonator and the noise in the system set the lower limit to the line width. Heterodyne measurements of the emission line width $\delta\nu$ of cw semiconductor lasers at low temperatures have given $\delta\nu = 150$ kHz for GaAs [53] and $\delta\nu \approx 54$ kHz for PbSnTe [54]. In the pulsed operation the p–n junction in a laser is surrounded by a time-dependent temperature field generated by the evolution of heat in the series active resistance of the laser crystal. We can assume that there are no nonradiative transitions in the p–n junction itself. Consequently, the heating of the laser crystal causes the individual modes to drift during a pulse, in accordance with Eq. (21). Therefore, the width of a mode integrated with respect to time under pulse operation conditions is a function of the pulse duration. This broadening of the emission line may sometimes be so great that it can be recorded using standard diffraction-grating instruments.

CHAPTER II

EXPERIMENTAL METHOD

§1. Apparatus for Excitation of Injection Lasers and Recording of Emission Spectra

Since all the investigated radiation sources were operated in the pulsed regime (pulse duration ranged from 0.1 to 10 μsec), the emission spectra were determined by a synchronous detection method similar to that described in [55]. This method was modified by replacing a generator of high-power pulses based on a hydrogen thyratron with a small-size semiconductor pulse generator [56], which was essentially a current aplifier (emitter follower). This generator was composed of two high-power transistors of the 2T803A type and it could deliver a power of ~1 kW (a current of about 30 A) to an active load of 1 Ω.

The circuit for the excitation of semiconductor lasers and recording their spectra is shown in Fig. 5. A pulse from a master generator of the 26I type was applied to the input of the semiconductor generator, whose load was the investigated laser diode. The semiconductor generator (amplifier) was supplied from a stabilized rectifier of the VS-12 type. The current through the investigated diode was measured with a vacuum-tube voltmeter of the V4-3 type, which measured the voltage drop across a 0.1 Ω resistor connected in series with the laser diode.

Depending on the required operation regime and the working temperature of the laser diode, it was excited with pulses of 0.5 to 20 A amplitude and of 0.3 to 10 μsec duraction. The repetition frequency ranged from 10^2 to 10^3 Hz.

Spontaneous or coherent radiation emitted by the laser diode was focused by a lens or an objective onto the entry slit of an IKM-1 grating monochromator. The radiation was then collected by a suitable lens or mirror and focused on the sensitive area of a photodetector. Different lenses and detectors were used, depending on the radiation wavelength. In the case of GaAs diodes we used an objective of the Yupiter-9 type, a quartz lens, and an FÉU-22 photomultiplier; in the case of InAs and InSb diodes, we used an LiF lens and InSb photodetectors; in the case of PbSe diodes, CO_2 laser, and InSb Raman laser, we used BaF_2 and KRS-5 lenses and Ge:Zn and Ge:Cu photodetectors. The BaF_2 and KRS-5 lenses were designed so that for a given focal length and $\lambda = 10.6$ μ, the spherical aberration was minimized and coma was eliminated. This made it possible to focus the CO_2 laser radiation into a spot of 200–300 μ size, which was particularly important in the excitation of the Raman lasers.

Fig. 5. System used for the excitation of injection lasers and recording of their emission spectra.

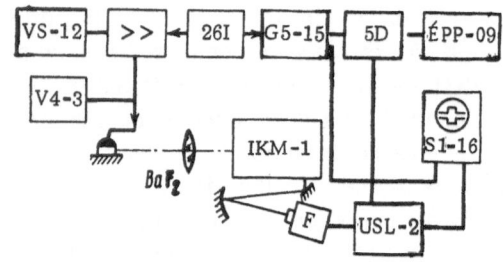

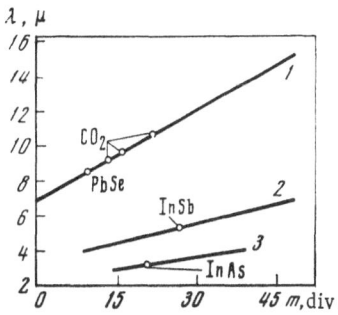

Fig. 6. Calibration graph of the IKM-1 monochromator. The numbers alongside the curve
give the diffraction order of the echelette grating;
circular points give the emission wavelengths of
the investigated lasers.

The detector signal was amplified with a USh-2 wide-band amplifier and applied to the
input of an SD synchronous detector. This detector received also a reference signal from a
G5-15 generator, which was triggered by the 26I master generator when injection lasers were
investigated and by a photodiode located near the rotating laser mirror when the CO_2 laser
was used. The emission spectra were recorded automatically using a potentiometer of the
ÉPP-09 or KSP-4 type. The circuit was checked with an S1-16 double-beam oscillograph.

The emission wavelengths of the investigated lasers were in the range 0.8-12 μ. The
mode structure of the semiconductor laser radiation or the rotational structure of the CO_2
laser radiation could be studied with a spectroscopic instrument providing a resolution of
~ 1 cm^{-1} at λ = 10 μ, i.e., a grating spectrometer was needed. The available IKM-1 infrared
monochromtor could be operated with gratings only up to 6 μ. *

In order to record the emission spectra of pulse light sources in the 3-15 μ range with
a sufficiently high resolution, we extended the working range of the IKM-1 monochromator by
using an echelette, bandpass filters, and a Ge:Zn photodetector [29].

The echelette had 100 lines/mm and its blaze angle was 30°; the ruled area was 57 × 57
mm. The spectral range from 3 to 15 μ could be covered by using the echelette in the first,
second, and third order (Fig. 6). During calibration the radiation of the required order was
selected with bandpass interference filters (in the 3-8 μ range) and with an absorption cutoff
filter (in the range of wavelengths exceeding 8 μ). The cutoff filter was an InSb plate $\lesssim$ 1 mm
thick whose transmission in the 8-15 μ range was 35-50% after the application of an antireflection coating.

The monochromator was adjusted using a Q-switched CO_2 gas laser which, under certain
conditions, generated three groups of lines at 9.2, 9.6, and 10.6 μ (spectra d, e, and f in Fig.
7). A fine calibration of the monochromator was made using the absorption lines of CH_4, CO_2,
NH_3, and water vapor. A Nernst rod was used as the radiation source in this calibration. The
absorption spectra were recorded using a narrow-band Tor-B system.

The calibration graph of the monochromator (Fig. 6) consisted of three lines, corresponding to the three echelette orders. The calibration lines were slightly nonlinear because
of the dispersion of the NaCl prism. Their average slopes were 16, 7.0, and 4.8 Å per one
small division of the drum for the first, second, and third orders, respectively.

Figure 7 shows the emission spectra of InAs, InSb, and PbSe semiconductor lasers and
of the Q-switched CO_2 gas laser, which were recorded using this monochromator. The mode
structure of the semiconductor laser radiation and the rotational structure of the CO_2 laser
radiation were clearly visible.

---

*Recently the Soviet industry produced a batch of prototype infrared grating spectrometers
  (IKS-16) with a working range up to 18 μ. However, this type of spectrometer is unsuitable
  for pulse measurements.

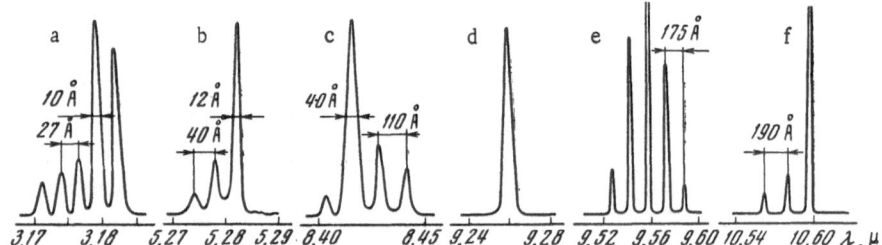

Fig. 7. Emission spectra of semiconductor lasers: a) InAs at 77°K;
b) InSb at 4.2°K; c) PbSe at 25°K; d, e, f) Q-switched CO_2 gas laser
at room temperature.

The minimum energy interval which could be resolved by the monochromator at the laser
emission wavelengths λ = 3.2, 5.2, 8.5, and 12 μ was about 6×10^{-5}, 5×10^{-5}, 7×10^{-5}, and
3×10^{-5} eV. The relative changes in the positions of the individual laser modes under the
influence of external agencies could be measured to within an error several times smaller.

A monochromator of the IKM-1 type was also used in recording the emission spectra
of p–n junctions in gallium arsenide. In this case the dispersive element was a diffraction
grating with 600 lines/mm, which concentrated light in the second order in the region of 0.7-1.1
μ. The monochromator was calibrated using the emission lines of argon.

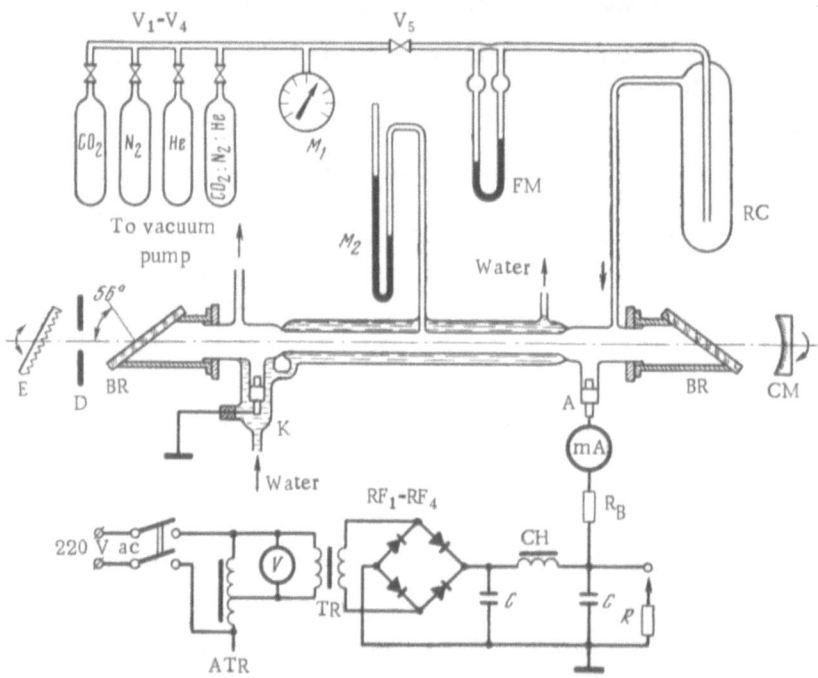

Fig. 8. Electrical, optical, and gas systems of a Q-switched CO_2 laser: ATR) RNO-250-5 autotransformer; TR) step-up transformer
(step-up ratio 50); RF_1-RF_4) D1009A rectifying elements; C) 4 μF
capacitors; CH) choke (8.7 H); R_B = 50 kΩ) ballast resistor; R = 10
kΩ) limiting resistor of a discharger; A) anode; K) water-cooled cathode; E) echelette grating (100 lines/mm); D) iris diaphragm; BR) rotatable Brewster NaCl plates; CM) concave mirror (radius of curvature 10 m); V_1-V_4) needle valves; V_5) fine-control needle valve; M_1)
standard manometer (100 kg/cm^2); M_2) oil manometer; FM) flow meter;
RC) reservoir cylinder (capacity 5 liters).

The energy resolution of this monochromator was $\sim 5 \times 10^{-4}$ eV at $\lambda = 0.81\ \mu$, which was insufficient for the resolution of the mode structure of the GaAs laser radiation.

§ 2. Q-Switched CO$_2$ Laser

The Raman scattering of light by free carriers localized in Landau levels was studied using a CO$_2$ gas laser emitting at 10 μ. A high output power was obtained by using a mixture of CO$_2$, N$_2$, and Ne, which was pumped slowly through a discharge tube. Since the lifetime of the vibrational levels of the CO$_2$ molecule was very long ($\sim 10^{-3}$ sec), the output power could be increased still further by modulating the Q factor of the resonator.

The electrical, optical, and gas systems of the carbon dioxide laser are shown in Fig. 8. A dc discharge was produced in a tube made of transparent quartz. The internal diameter of the the tube was 27 mm and its length was 2 m. A high-voltage rectifier with an LC filter was used as the supply source. The voltage across the rectifier output could be varied from zero to 12.5 kV and the current of up to 200 mA could be supplied to the load. The tube and the cathode K were cooled with running water. The tube was closed with two sodium chloride plates oriented symmetrically at the Brewster angle. These plates ensured a high degree of polarization of the laser radiation ($\sim 99\%$) [57] and their symmetric positions prevented displacement of the beam. The plates were bonded with an epoxy resin to holders which could be rotated through 360°K without disturbing the vacuum in the tube; in this way any position of the plane of polarization could be selected by suitable rotation. For safety reasons the anode holder was made of Plexiglas. This construction made it easy to replace the resonator.

A selective resonator consisted of a concave mirror M with a radius of curvature of 10 m and a reflecting echelette E with 100 lines/mm and a blaze angle of 30°. The mirror was coated with gold and placed in a holder which was fitted to an axle of a DID-1TA electric motor. When the motor axle was carefully balanced, the mirror could be rotated at several hundreds of revolutions per second. The rate of rotation was selected so as to ensure minimum beats of the axle. The echelette was used in an autocollimation configuration with the energy concentrated in the zeroth order. The echelette was attached to a rotatable table so that the resonator could be tuned to any one emission line of the vibration-rotational spectrum of the CO$_2$ molecule at 9.2, 9.6, 10.2, and 10.6 μ. An iris diaphragm D placed inside the resonator made it possible to select the lowest order transverse mode (this was checked by examining the far-field radiation pattern of the laser when it was operated continuously). Moreover, this diaphragm made it easier to suppress the generation of the neighboring rotational lines, which could occur because of the competition between the lines since the pulse duration ($\sim 10^{-7}$ sec) was comparable with the rotational level lifetime.

The mixture of CO$_2$, N$_2$, and He was prepared in one of the cylinders (Fig. 8) using valves V_1-V_4 and a manometer M_1. A fine-control needle valve V_5 was used to pass the mixture to a reservoir cylinder RC in order to suppress the gas pressure pulsations. It was then pumped through the discharge tube with a VN-2MG vacuum pump. The flow of gas was measured with a flow meter FM. The total pressure in the mixture in the middle of the discharge tube was measured with a U-type oil manometer M_2.

The output power and the emission spectrum of the laser depended mainly on the type of resonator, operating condition, current through the tube, and composition and pressure of the gas mixture in the tube.

Under continuous operating conditions, when the echelette was replaced with a concave mirror of radius 6 m and a coupling aperture of 5 mm, we found that the output power was about 80 W at $\lambda = 10.6\ \mu$. When the Q factor of this resonator was switched, the output pulses were of 1-2 μsec duration and 7-8 kW power. The emission spectrum then consisted of several groups of lines at 9.2, 9.6, and 10.6 μ and these were of different intensities (Fig. 7).

The separation between the lines was ~56 GHz. The strongest group of lines formed a band at 10.6 μ. The intensity of the 9.6 μ band was an order of magnitude lower. The other bands were even weaker.

A selective resonator made it possible to vary the output power of individual lines from several watts to several tens of watts. This power was highest when the plane of polarization was parallel to the echelette lines. In this case the switching of the Q factor produced pulses of ~0.3 μsec duration and ~1 kW power. The emission spectrum consisted practically of one line (see Fig. 42 in Chap. IV). The laser usually emitted the P_{20} line (due to the 00°1–10°0 transition) at $\lambda = 10.5915$ μ. Typical operation conditions were as follows: the current was 30–40 mA, the voltage drop across the tube was 8 kV, the composition of the gas mixture was CO_2:N_2:He = 1.5:1.5:10, the gas flow was 0.5 liter/min, the pressure in the gas mixture was ~10 torr, and the rate of rotation of the mirror was ~200 rev/sec.

§3. Technique Used in Low-Temperature Magnetooptic

Investigations at Infrared Wavelengths

Magnetic fields were generated by two magnets, one of which was a conventional electromagnet producing up to 13 kOe and the other one was a superconducting solenoid producing fields up to 56 kOe. Helium optical cryostats of 1.3–2.2 liter capacity of two types were used in conjunction with the conventional electromagnet: a sample was placed on a heat sink in a helium bath or it was immersed directly in the cryogenic liquid. A special optical cryostat of cylindrical type [58] was constructed for use in conjunction with the superconducting solenoid.

A. Apparatus Used in Fields up to 13 kOe. The electromagnet consisted of a yoke made of Armco iron and of 16 water-cooled sections. These sections were made of a copper strip of the PBD type (dimensions 10×1.25 mm) and they were impregnated with bakelite lacquer. The electromagnet was placed on a rotatable table. The gap could be varied by changing demountable pole-pieces made of Armco iron or of Permendur. The maximum field intensity was ~10 kOe (for a current of ~19 A) when the gap was 41 mm and the magnet winding was supplied with current from a VSA-5 rectifier. When an ÉMU-100 amplidyne was used as the power source, the magnetic field in the gap reached 13 kOe for a current of 42 A. The magnet was calibrated using an IMI-3 magnetic induction meter.

The optical cryostats used in conjunction with this magnet had long side branches with an external diameter of 40 mm. Two recesses at the ends of these branches were used for the attachment (with an epoxy resin) of "warm" BaF_2 windows, which were transparent up to $\lambda = 15$ μ. When a sample was immersed in liquid helium or liquid nitrogen, internal "cold" windows were held in place by flanges and had indium seals. In the wavelength range up to 6 μ the cold windows were made of sapphire and in the range up to 8.5 μ these windows were made of a silicon single crystal which was antireflection-coated to give a transmission of 90%. Care in the making of the indium seal of the silicon windows ensured that they operated reliably at nitrogen and helium temperatures. The aperture angle of the cryostats was ~80°.

B. Apparatus Used in Fields up to 56 kOe. In this case the magnet was a one-section solenoid which produced magnetic fields up to 56 kOe. The working diameter inside the solenoid was 31 mm and the critical current was 112 A. The solenoid was wound using a superconducting partly stabilized cable of the KSI-042/0.07 type composed of six wires (0.27 mm diameter) of the superconducting alloy BT-65 and one copper wire, which was used for stabilization [59]. The cable wires were twisted and impregnated with indium. The cable was insulated with a polyethylene terephthalate filament. The solenoid winding consisted of 52 layers; the total number of turns was 7137 and the length of the cable was 2050 m. The space factor of the cable was 0.79. The ratio of the length of the winding $l = 145$ mm to the average diameter d_{av} was 1.6. The calculated homogeneity of the field [60] was 1% in a sphere of 13.5 mm radius

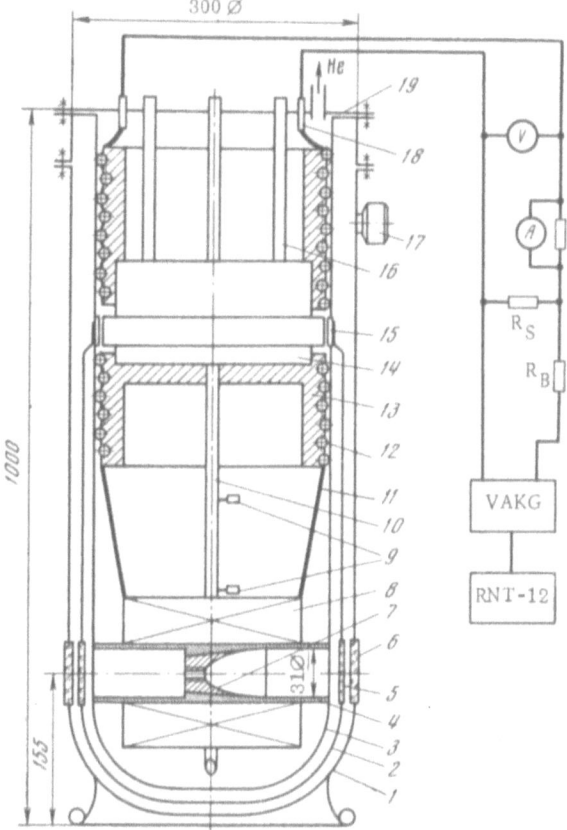

Fig. 9. Schematic cross section through a cryostat and power supply to a superconducting solenoid: 1) frame; 2) screen; 3) helium bath; 4) copper tube with conical sections; 5) thermal filter; 6) NaCl or BaF_2 window; 7) parabolic mirror; 8) solenoid; 9) carbon resistor; 10) tube for pouring in helium; 11) Pb leads of 12.5 mm cross section; 12) current leads; 13) plastic foam; 14) nitrogen bath; 15) copper ring; 16) tubes for puring in nitrogen; 17) tube for pumping out cryostat; 18) copper terminals; 19) upper flange; VAKG) rectifier; RNT-12) autotransformer; R_S) external shunt of solenoid; R_B) ballast resistor.

and 0.05% in the sphere of 3 mm radius. The solenoid frame was made of anodized Duralumin. The cooling of the winding was improved by making apertures of 4 mm diameter in the frame ends. The surface of the frame which was in contact with the winding was covered with two layers of a polyethylene terephthalate film 20 μ thick and these layers were bonded with a BF-2 adhesive. The winding layers were separated by the same type of film.

The magnet was supplied from a VAKG three-phase rectifier. When the oscillation frequency was 300 Hz and the solenoid inductance was 1.3 H, the alternating component of the current could be ignored. The current was controlled with an autotrasformer connected across the rectifier output and a ballast resistance R , connected in series with the solenoid (Fig. 9). An external shunt resistance $R_S = 0.25$ Ω ensured that the energy evolved as a result of

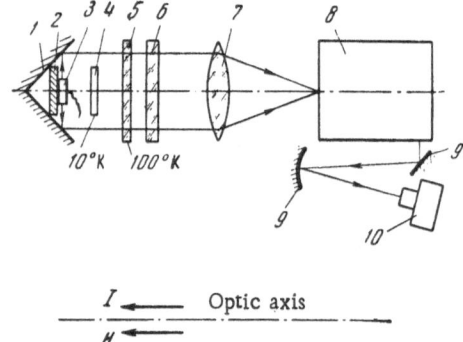

Fig. 10. Optical system for recording radiation emitted by injection lasers in a longitudinal magnetic field produced by a superconducting solenoid: 1) bimirror; 2) copper heat sink; 3) laser; 4) copper screen; 5) thermal filter; 6) window; 7) lens; 8) monochromator; 9) focusing mirrors; 10) photodetector. The lower part of the figure shows the directions of the injection current and magnetic field relative to the optic axis.

loss of superconductivity by the solenoid was dissipated in 5-6 sec. This dissipation time of magnetic energies up to 8 kJ ensured that no damage occurred and that only ~2 liters of helium was lost by evaporation.

In the course of calibration of the solenoid the magnetic field was measured by a ballistic method and with InAs and InSb Hall-effect probes.

A cylindrical metallic cryostat with a low-temperature working chamber and optical access to this chamber was used in conjunction with the superconducting solenoid. The cryostat construction ensured efficient use of the space inside the solenoid. The internal diameter of the cryostat was 260 mm. It is shown schematically in Fig. 9. A solenoid 8 was wound on a copper tube 4, which was soldered with tin into a helium bath 3. The copper tube ensured optical access to the working chamber of the solenoid through demountable* windows 6 and demountable filters 5; it also acted as a heat sink for the investigated sample. The liquid-helium cooling of the internal part of the solenoid and of the copper tube was improved by a uniform annular gap between them, which was 1 mm wide. Various optical elements were placed in a conical section (10° angle) inside the copper tube 4. In the measurements of the emission spectra of the injection lasers in a longitudinal magnetic field we used a bimirror with a conical holder (Fig. 10), whereas in studies of the Raman scattering by Landau levels we employed a parabolic mirror (in the spontaneous scattering case) or a conical holder with two rotatable mirrors (in the stimulated scattering case). The bimirror and the rotatable mirrors were optically polished Duralumin plates coated with aluminum films. The parabolic mirror was made of copper† and coated with gold. A threaded aperture was made at the center of this mirror: it was intended for screwing in a copper sample holder.

This threaded aperture, conical sections, and other joints of all the parts of devices inside the tube were coated with silicone grease, which had a high thermal conductivity so that the temperature of the sample holder did not exceed 10°K.

A helium bath was made of stainless-steel sheet 0.8 mm thick. The bath was surrounded by a copper screen 2 (Fig. 9), which was of 280 mm diameter and had walls 1 mm thick; further screening was provided by a nitrogen bath 14. The screen was cooled by the evaporating helium, which passed through an annular gap between the nitrogen bath and a copper ring 15. Under these conditions the screen temperature was ~100°K. Copper tubes used as current leads (12) were of 6 mm diameter and had 1 mm walls. They were surrounded by a plastic foam 13 in such a way that one of them was pressed tightly against the helium bath 3 and the

---

*The use of demountable windows and filters in the cryostat emabled us to modify it quite easily for use with a "warm" horizontal aperture of 20 mm diameter.

† This mirror was made using a cutter with a parabolic profile with F = 2 mm. The profile was obtained by cutting a cone with a plane parallel to the cone generator.

Fig. 11. External appearance of an optical cryostat with a superconducting magnet.

other formed a helical gap with this wall. Evaporating helium passed along this gap and it carried away the heat from both copper tubes [61]. These tubes (12) were insulated electrically by a polyethylene terephthalate film. The purpose of the other parts of the cryostat is explained in Fig. 9.

Not more than 6 liters of helium was lost in the cooling of the solenoid and helium bath, which were precooled to liquid nitrogen temperature. The amount of helium needed inside the cryostat was 13-18 liters and its rate of evaporation was 0.4 liter/h.

In the infrared range we used NaCl windows. Germanium plates, with coatings preventing reflection in the investigated part of the spectrum, were used as thermal filters.

The external appearance of the cryostat is shown in Fig. 11.

§ 4. Apparatus Used in Optical Measurements at Infrared

Wavelengths under High Hydrostatic Pressures at 77°K

The high-pressure apparatus was developed and made at the Institute of the Physics of High Pressures of the Academy of Sciences of the USSR. A description of this apparatus was given in [62-66].

We shall briefly review the features of the apparatus which provided high hydrostatic and quasihydrostatic pressures and we shall consider in greater detail the optical parts which were developed with the cooperation of the present author [66].

Usually, the high-pressure investigations were carried out in a medium which transmitted pressure hydrostatically to a sample. Since at liquid nitrogen temperature the majority of the gases solidified at high pressures, we selected helium as the pressure-transmitting medium. Helium solidified at 77°K when the pressure was 14.15 kbar.

A schematic diagram of the apparatus used to study lasers under high hydrostatic pressures at 77°K is given in Fig. 12. A compressor was used to raise the helium pressure to 2 kbar and further compression up to 5 kbar was obtained with a pressure booster, which was connected by a capillary made of the 1Kh18N9T steel to an optical chamber 3. This chamber was placed at the bottom of a plastic-foam container 6. The working volume of the chamber was of 10 mm diameter and 10 mm long. A diode 4 was placed inside this chamber on an insulating holder. The diode radiation passed through a conical window 5 and a vacuum cell 7, and it was focused by a lens 9 onto the entry slit of a monochromator. The other parts of the apparatus are explained in Fig. 12.

The main feature of the apparatus used in the present investigation, which distinguished it from that described in [67], was the presence of a vacuum cell. This cell allowed us to use a plastic-foam Dewar container rather than an evacuated one, which simplified the use of the high-pressure chamber.

The vacuum cell was a thin-walled tube made of Textolite. Since the thermal conductivity of Textolite was low, the outer end of the cell was kept at room temperature and it carried a sodium chloride window. During investigations the cell was evacuated with a backing pump and the window was kept warm by passing hot air around it. The cell diameter was selected so as to avoid reducing the maximum aperture angle, which was governed by the construction of the conical window. This angle was 14°.

Pressure-contact windows were usually employed in the infrared range [19, 68]. These windows were made of sapphire, diamond, polycrystalline germanium or silicon with a grain size of ~1 μ; the window thickness was approximately equal to its diameter. Use was also made of conical windows [66], which were made of germanium, silicon, and sapphire single crystals. The advantage of the conical window was the reliability of the seal. Moreover, the internal pressure pushed the window into the conical socket and this gave rise to a supporting pressure along the generator of the cone, which was a desirable feature. Furthermore, the amount of the material required to produce a conical window with a given aperture was considerably smaller than the amount required for a conventional window.

The silicon windows were more reliable than germanium windows. At hydrostatic pressures up to 5 kbar, a silicon window withstood repeated rises of the pressure at rates up to

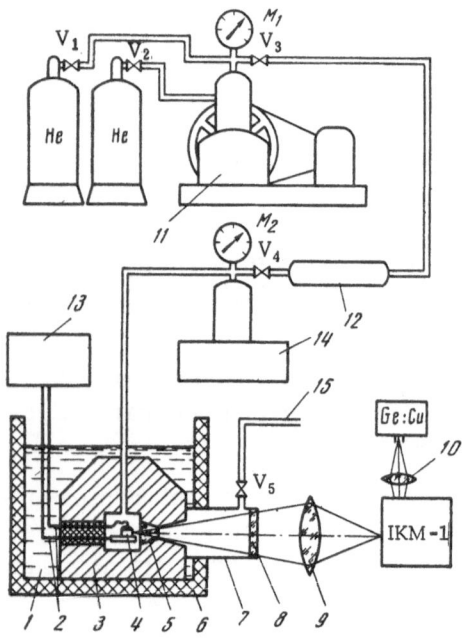

Fig. 12. Schematic diagram of the apparatus used to generate high hydrostatic pressures in an optical chamber at 77°K: 1) liquid nitrogen; 2) electrical leads; 3) high-pressure chamber; 4) laser; 5) Ge, Si, or sapphire windows; 6) plastic foam; 7) cell made of Textolite; 8) NaCl or quartz windows; 9) BaF$_2$ lens (or Yupiter-9 objective); 10) NaCl or quartz lens; 11) compressor; 12) filter; 13) power supply to laser; 14) pressure booster; 15) connection to vacuum pump; M$_1$ and M$_2$) manometers; V$_1$–V$_4$) high-pressure valves; V$_5$) vacuum valve.

10 bar/sec and repeated falls of the pressure at rates up to 100 bar/sec without disturbance of the seal and without loss of the optical transparency.

The apparatus described above was used at hydrostatic pressures up to 5 kbar at T = 77°K mainly in investigations of the PbSe laser radiation. We also used apparatus which produced quasihydrostatic pressures up to 10 kbar at T = 77°K [65], which was employed in studies of the GaAs laser radiation. In the latter case the pressure-transmitting medium was a mixture of oil with kerosene. Sapphire windows were of the pressure type. The pressure in the chamber was established at room temperature. Then, the chamber was placed in a glass Dewar container and was slowly (in 4 h) cooled to liquid-nitrogen temperature.

The pressure coefficient of the forbidden-band width of GaAs (up to 2.7 kbar) was determined at 77°K using helium as the pressure-transmitting medium. The local pressure in the medium was determined by measuring the laser radiation wavelength when the oil and kerosene mixture was frozen. The absolute error in this determination of the pressure was ~500 bar when the pressure was 9 kbar because the calibration curve was obtained by extended extrapolation from low pressures (~2 kbar). The error in the measurement of the relative change in the pressure was an order of magnitude smaller and it resulted from the error in the measurement of the laser wavelength ($\delta\lambda \approx 3$ Å, which corresponded to $\Delta P = 50$ bar). This method allowed us to measure the pressure at points in the frozen medium separated by distances up to 13 mm from the optic axis of the chamber and this was done at $P \approx 6.7$ kbar. The laser radiation was extracted from the high-pressure chamber with a rotatable mirror. There was no pressure gradient (within the experimental error of ~50 bar). This circumstance as well as the fact that GaAs diodes did not fracture under pressure, led us to conclude that the medium used up to 9 kbar at 77°K produced hydrostatic pressure.

Standard SVD high-pressure manometers with upper limits of 2.5×10^3, 4×10^3, 6×10^3, and 10×10^3 kg/cm^2 were used in pressure measurements. The absolute values of the pressure were determined to 1%. The construction of these manometers was such that the relative change in the pressure could be determined with a higher precision (the error did not exceed 30 kg/cm^2 when a manometer with an upper limit of 10^4 kg/cm^2 was used). In some measurements we used a magnanin resistance manometer with $R_0 = 100$ Ω. In this case the pressure was determined to within 20 bar.

§ 5. Zinc- and Copper-Doped Germanium Infrared-Radiation

Detectors

The radiation emitted by PbSe diodes, CO$_2$ laser, and Raman laser was detected using germanium doped with zinc (sensitivity range 1.6–14.5 μ) and germanium doped with copper (sensitivity range 8–30 μ). The zinc-doped detector was operated at liquid-nitrogen temperature (~55°K) and the copper-doped detector was used at liquid hydrogen temperature (~25°K). The latter (Ge:Cu) detector could be operated also at liquid helium temperature. However, we preferred the use of liquid hydrogen because the resistance of the detector was still sufficiently high (~10^4 Ω) at liquid hydrogen temperature. Moreover, a helium cryostat of the KR-15 type (of 1.5 liter capacity) could retain hydrogen in the liquid state for a week and this was a very convenient practical feature.

We shall now consider the main characteristics and construction of our detectors.

A. Ge:Zn Detector. Zinc was known to behave as a doubly charged acceptor impurity in germanium with two levels at 0.035 and 0.095 eV. Therefore, in order to use this detector in the 10 μ range, it was necessary to compensate the lower level with a donor impurity in such a way that $N_{zn} \leq N_D < 2N_{zn}$, where N_{zn} is the zinc concentration and N_D is the concentration of the donor impurity (Sb, As).

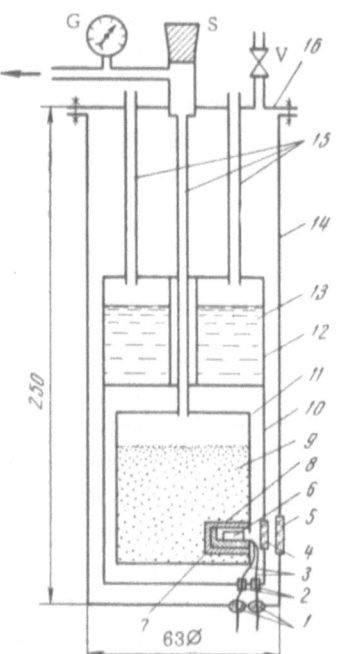

Fig. 13. Schematic diagram showing a Ge:Zn detector operating at solid-nitrogen temperature: 1) beads; 2) insulators; 3) manganin leads of 0.05 mm diameter; 4) Ge window coated to prevent reflection at 10 μ; 5) BaF$_2$ window; 6) sensitive element; 7) sapphire substrate; 8) copper cartridge; 9) solid nitrogen; 10) nitrogen screen; 11) solid nitrogen container; 12) liquid nitrogen bath; 13) liquid nitrogen; 14) casing; 15) stainless-steel tubes; 16) cover; V) valve for pumping out cryostat; S) stopper; G) standard VOSh-160 vacuum gauge.

The sensitive element of the detector was prepared from germanium doped with zinc to give $N_{zn} \approx 10^{16}$ cm^{-3} and partly compensated with antimony ($\sim 1\%$). The sensitive area was 2×2 mm. The length of the sample was $l = 1$ cm, so that the condition $\alpha l > 1$ was satisfied where $\alpha \approx 4$ cm^{-1} was the absorption coefficient of light for the 0.095 eV level at the impurity concentration employed [69].

The ends of the detector crystal were optically polished. Indium contacts were alloyed to two opposite faces and then manganin wires of 0.05 mm diameter were soldered to the contacts. One of the free surfaces was coated with an epoxy resin and bonded to a sapphire substrate, which had a very high thermal conductivity at liquid nitrogen temperature. The detector was then placed, with indium spacers and silicone grease, in a copper cartridge 8 mounted inside a small helium cryostat (Fig. 13).

When this cryostat was assembled and evacuated, both chambers 11 and 12 were filled with liquid nitrogen and nitrogen vapor was pumped out of the inner chamber 11 until the pressure fell to several torr. At these pressures the nitrogen solidified and the temperature of the sample was $\sim 55°$K. Figure 14 shows the dependence of the resistance of a Ge:Zn detector element on the pumping duration. This dependence reached saturation in about 1 h and the resistance then remained quite stable for about one working day.

The detector parameters were measured at the Scientific-Research Institute of Applied Physics. For the best element these parameters were as follows: integrated responsivity

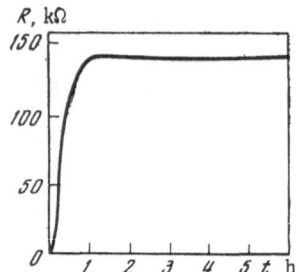

Fig. 14. Dependence of the resistance of a Ge:Zn detector on pumping duration. R(300°K) = 15 Ω; R(80°K) = 220 Ω; R(55°K) = 140 kΩ.

Fig. 15. Mounting of two Ge:Cu detectors in a branch tube of a helium cryostat; 1) manganin leads of 0.05 mm diameter; 2) beads; 3) insulators; 4) germanium window coated to prevent reflection at 10 μ; 5) window; 6) diaphragm; 7) sensitive element; 8) sapphire substrate; 9) copper cartridge; 10) helium screen; 11) helium bath; 12) nitrogen screen; 13) jacket; 14) liquid helium (or hydrogen).

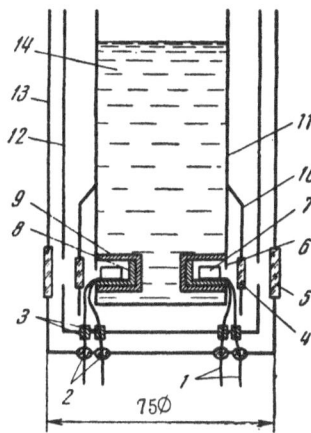

7600 V/W, detectivity 2.5×10^9 cm $\cdot$ Hz$^{1/2} \cdot$ W^{-1}. When the load resistance was 75 Ω the time resolution of the detector was 0.3 μsec. The angular size of the viewing field of the detector was 30°.

B. Ge:Cu Detector. Copper was known to form three acceptor levels in germanium. One of these levels ($\mathscr{E}_A \approx 0.04$ eV) was used to detect radiation in the middle infrared range. Although the maximum responsivity of the Ge:Cu detector was located at 26 μ, it was still one of the best detectors for signals in the 10 μ range [70].

This type of detector was prepared from materials of two kinds: 1) germanium with a copper concentration of $\sim 10^{15}$ cm^{-3} and compensated with up to 5% of antimony; 2) germanium with a copper concentration of 1.5×10^{15} cm^{-3} and compensated with 2% of antimony. The detectors prepared from the second material had a higher responsivity.

The detector element was a parallelepiped with a sensitive area of 3×3 mm and it was 6 mm long. This length was sufficient because the absorption cross section of copper was twice as high as that of zinc [69]. The procedure used in the preparation and mounting of the element was the same as in the case of the Ge:Zn detectors, except that in the present case a conventional helium cryostat (15) was used. The various parts of this cryostat are explained in Fig. 15.

The detector resistance at ~ 25°K was about 40 kΩ and the signal voltage produced in the 10 μ range was four times as high as that produced by the Ge:Zn detector; the time constant was $\lesssim 10^{-7}$ sec. The angular aperture of the viewing field of the detector was ~ 30°.

Figure 16 shows the shape of a radiation pulse emitted by a Q-switched CO$_2$ laser, recorded using a detector load R $= 75 \Omega$. Pulses of this type were obtained using either a Ge:Cu or a Ge:Zn detector. The duration of the radiation pulses was about 0.3 μsec.

The experiments under high pressures and studies of the magnetooptic effects were fairly difficult to perform and the availability of two types of detector (Ge:Zn and Ge:Cu) made

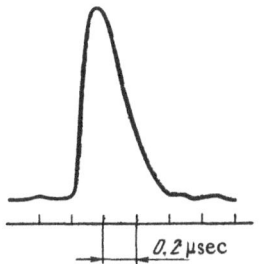

0.2 μsec

Fig. 16. Shape of radiation pulses emitted by a Q-switched CO$_2$ laser.

the task much easier. Nevertheless, in the Raman scattering experiments it was sometimes necessary to record simultaneously the useful signal and the pump pulse. When a Ge:Cu detector was used, the two signals could be observed simultaneously if two separate sensitive elements were employed (Fig. 15).

§6. Scanning of Infrared Radiation Emitted from

InSb Crystals

It is difficult ot investigate the spatial distribution of infrared radiation. In studies of electroluminescent diodes there is an additional difficulty because of their small size ($\lesssim$ 0.5 mm). The currently available infrared microscopes can be used only up to $\lambda = 1.3\ \mu$.

In view of this situation the recombination radiation emitted from InSb diodes was studied using a Cassegrain-type mirror objective. The advantages of this objective included achromaticity so that it could be easily adjusted using visible light. The resolution of this objective in the $\lambda > 4\ \mu$ range was governed by the diffraction divergence

$$\Delta l = 0.61 \lambda / A, \tag{23}$$

where A is the aperture (A = 0.116) and Δl is the minimum resolvable linear element. For $\lambda = 5.2\ \mu$, we found that $\Delta l \approx 27\ \mu$.

A system for the scanning of the spontaneous radiation emitted from a sample is shown in Fig. 17. The investigated face of a sample 1 was magnified with an objective 3 by a factor of about 30 and the image was focused (in the visible range) on a screen 5. The screen was then removed and replaced with an InSb photodetector (not larger than 0.8 mm) and this detector was moved by a mechanical X-Y drive in the image plane. The system could be automated by fixing the photodetector position and by making a rotatable mirror 4 execute oscillatory motion in two mutually perpendicular planes.

In those cases where specific location had to be known (for example, the position of the emitting region relative to the contacts), it was necessary to employ a structure composed of two diodes and shown in the inset in Fig. 17. Here, an InSb diode 7 and a GaAs diode 9 were soldered with their p-type regions to a copper strip 8 in such a way that a distance from the p–n junction in GaAs to the InSb crystal was about 0.2 mm, checked separately with a microscope. Since the emission region of the GaAs diode was only 10 μ and the InSb detector was highly sensitive to the $\lambda \approx 0.85\ \mu$ radiation, the GaAs diode provided a good standard in such measurements.

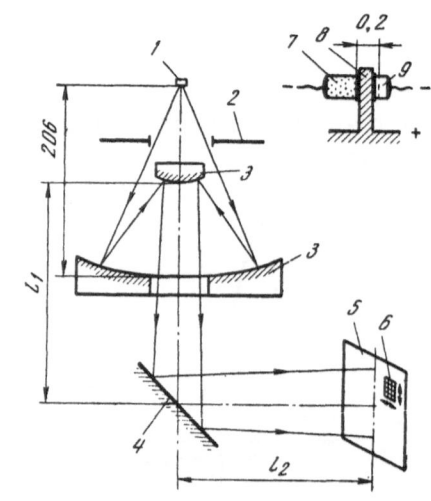

Fig. 17. System for scanning infrared radiation with a Cassegrain-type mirror objective; 1) sample; 2) diaphragm; 3) Cassegrain objective; 4) rotatable mirror; 5) image plane (screen); 6) sensitive area of a photodetector moved in the image plane by an X-Y drive mechanism; 7) InSb diode; 8) copper strip; 9) GaAs diode (standard); $l_1 + l_2 = 1600$ mm.

§7. Other Measurements

1. The average power of a Q-switched CO_2 laser, which amounted to hundreds of milliwatts, was measured with a spherical calorimeter. The absorbing load was a sphere (model of an absolute black body) made of copper. The internal surface of this sphere was coated electrolytically with a black layer. The absorption of radiation raised the temperature of this sphere and this temperature was measured with 15 Chromel–constantan thermocouples. The thermocouple emf's were determined with a KP-59 potentiometer. The calorimeter was calibrated by winding a heater on the sphere.

The minimum power which could be recorded with this calorimeter was 5×10^{-4} W. The responsivity of this calorimeter was 3.3×10^{-2} V/W. The time constant was 3 min.

2. The average output power of semiconductor lasers was measured with a vacuum radiation thermoelement made at the Leningrad Electrical Engineering Institute. The responsivity of this element was 4.8 V/W. It consisted of two sensitive components (for the compensation of the background radiation during measurements) and they had sensitive areas of 1×1 mm. These components were enclosed in a glass cell with a BaF_2 window. The emf was measured with an M195/1 microammeter.

3. The polarization of the radiation was measured with an IPP-12 infrared polarizer, composed of eight AgCl plates. The degree of polarization achieved in this way was 95% and the transmission was 35-40%. The relative aperture of this polarizer was 1:4.5.

CHAPTER III

INFLUENCE OF MAGNETIC FIELDS ON EMISSION SPECTRA OF p−n JUNCTIONS IN InAs, InSb, AND PbSe

§1. Spontaneous and Coherent Radiation Emitted

from InAs Injection Lasers

A. Samples. Injection lasers with p−n junctions were fabricated by the diffusion of cadmium or zinc into n-type indium arsenide with an initial donor (Te) concentration n = $(2.5-3.0) \times 10^{18}$ cm^{-3} and with an electron mobility of $(8-10) \times 10^3$ cm$^2 \cdot$ V$^{-1} \cdot$ sec^{-1} at 300°K. A p−n junction was typically located at a depth of 20-50 μ. A Fabry–Perot resonator was obtained by cleaving along the (110) crystallographic planes at right-angles to the p−n junction plane. A film of gold was evaporated on the p- and n-type regions and contacts were indium-soldered to these regions. Diodes were placed in a copper corner holder.

Table 6 lists the more important parameters of the investigated diodes (l is the resonator length; a and h are the thickness and width of the crystal, respectively; S is the p−n junction area).

TABLE 6. Some Properties of InAs Diodes

Diode No.	n × 10^{-18}, cm^{-3}	Acceptor	l, mm	a, mm	h, mm	S, mm^2
427	2.5	Cd	0.46	0.25	0.2	0.11
527	2.5	Cd	0.50	0.28	0.2	0.14
2	3	Zn	0.50	0.40	0.2	0.20
3	3	Zn	0.50	0.34	0.2	0.17
4	3	Zn	0.45	0.16	0.2	0.07
5	3	Zn	0.56	0.45	0.2	0.25

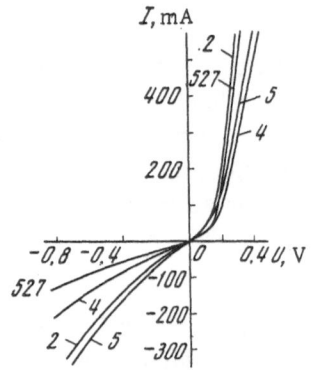

Fig. 18. Current — voltage characteristics of InAs diodes at 300°K. The diode numbers are given alongside each curve.

Static current — voltage characteristics of some of the InAs diodes were recorded at 300 and 77°K (Figs. 18 and 19, respectively).

The reverse current showed no saturation, which was typical of the zinc-doped diodes. At 77°K this was due to the tunneling of carriers to the density-of-states tails in the forbidden band or to other levels, and also due to surface leakage losses. The predominance of the tunneling mechanism was indicated by the observation that the initial reverse current observed in the range $U \le |0.15|$ V was usually higher than the forward current. At 300°K an additional contribution to the reverse current was made by the generation of carriers in the space-charge region [71].

The forward branches of the current — voltage characteristics were given by $I = I_0 \exp (eU/rk_0T)$, where $r \approx 2$ at 300°K and $r \approx 3-7$ at 77°K. The high values of r at 77°K indicated a large excess current. A comparison of the characteristics of two diodes (Nos. 4 and 5) with very different ratios of the p—n junction area to its perimeter indicated that the surface current was important. The cutoff voltage, found by extrapolation of the linear part of the characteristic, was 0.15-0.19 V at 300°K and 0.36-0.39 V at 77°K, which was slightly below the contact potential of the investigated material at these temperatues [72].

In the range of stronger bias voltages (U > 0.2 V at 300°K and U > 0.4 V at 77°K) the dependence $I = f(U)$ was linear. The slope of the linear part was governed by the resistance of the contacts and the bulk of the sample; this slope was 0.12-0.16 Ω at 77°K and 0.23-0.28 Ω at 300°K.

B. Weak Magnetic Fields. The influence of magnetic fields on the radiation emitted from InAs diodes was investigated at 4.2, 10, 30, and 77°K. It was found that the results depended strongly on the direction of the magnetic field relative to the current I through the diode [73].

The application of a transverse magnetic field ($\mathbf{I} \perp \mathbf{H}$) altered the radiation intensity, position of the maximum, width of the emission line, and threshold current density. These changes depended on the temperature. On the other hand, the application of a longitudinal

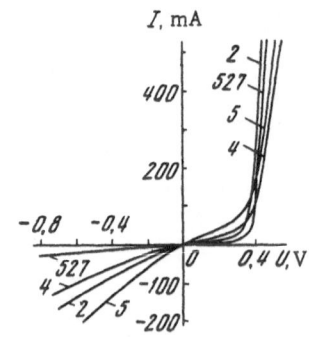

Fig. 19. Current — voltage characteristics of InAs diodes at 77°K. The diode numbers are given alongside each curve.

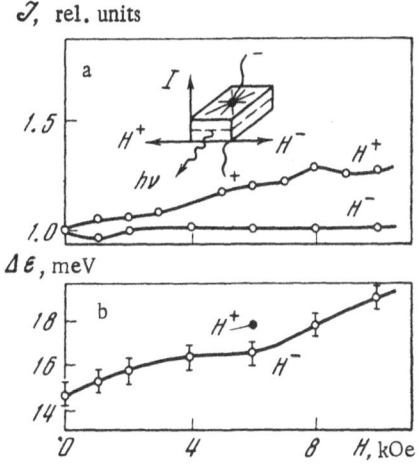

Fig. 20. Dependences of the intensity (a) and width (b) of a spontaneous line emitted by an InAs diode on the transverse magnetic field applied at liquid helium temperatures.

magnetic field ($I \parallel H$) caused no changes in the spectra of the spontaneous or coherent radiation at all the investigated temperatures (this was true within the limits of the resolving power of the apparatus).

Figure 20 shows the dependences of the spontaneous radiation intensity and line width on the transverse magnetic field applied at helium temperatures. The positive sign of H represents such a relative direction of the current and field that the Lorentz force deflects carriers toward the face from which radiation is emitted; the minus sign represents the opposite field. The top part of Fig. 20a shows the relative directions of the current, field, and emitted radiation. It is clear from this figure that the radiation intensity rises with the field H^+ and the line width increases for both directions of the field.

Moreover, we found that a transverse magnetic field shifted the spontaneous radiation maximum toward shorter wavelengths when the current through the diode was fixed (Fig. 21a). This shift depended on temperature and it amounted to ~ 4 meV for $H^+ = 10$ kOe, $I = 15$ A, and $T = 77°K$. A similar shift of the radiation in the direction of shorter wavelengths was observed also in a field $H = 0$ when the current through the diode was changed by a factor exceeding 5

Fig. 21. Energy position of the spontaneous radiation maximum plotted as a function of the magnitude and direction of the magnetic field for different currents through a diode (a) and as a function of the current through the diode (b) at temperatures of 4.2 and 77°K.

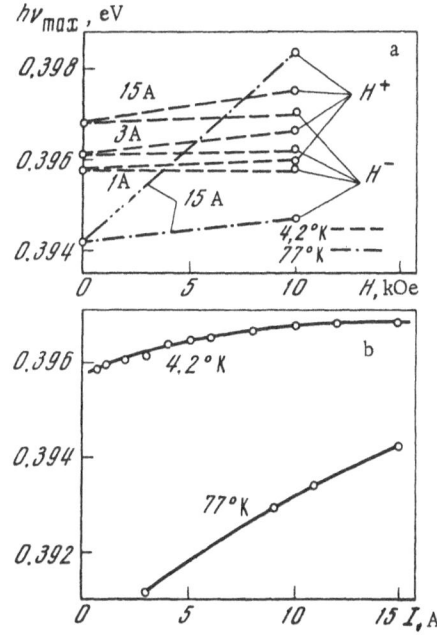

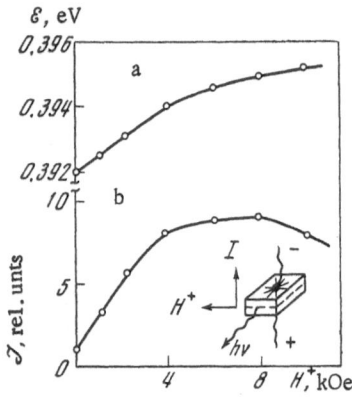

Fig. 22. Dependences of the center of gravity of the mode energy (a) and of the radiation intensity (b) on the transverse magnetic field applied to an InAs laser in the case when carriers are deflected by the field to the surface being examined. Diode No. 427 at $T \approx 10°K$; $j = 1.5 \times 10^3$ A/cm^2, $\tau = 1$ μsec.

(Fig. 21b). In this case the current pulses were made sufficiently short ($\lesssim 1$ μsec) in order to minimize the long-wavelength shift due to the heating of the diode. The shift of the maximum was small in fields H^-.

Similar dependences were also observed for coherent radiation but the effects were much stronger (Fig. 22). For example, in the coherent radiation case a field $H^+ \approx 6$ kOe increased the integrated intensity by a factor of 9.

Moreover, a transverse magnetic field reduced strongly the threshold current density. Figure 23 shows the lux−ampere characteristics of a diode No. 527 obtained at 77°K in different magnetic fields. Region II in these characteristics corresponds to the beginning of the laser action. It is clear that an increase of the field up to 10 kOe halved the threshold current. Similar results were obtained when the threshold was deduced from the emission spectra.

Figure 24 shows the dependence of the threshold current density on the magnetic field obtained for the same diode at temperatures of ~10 and 77°K. Cooling to 10°K resulted in a

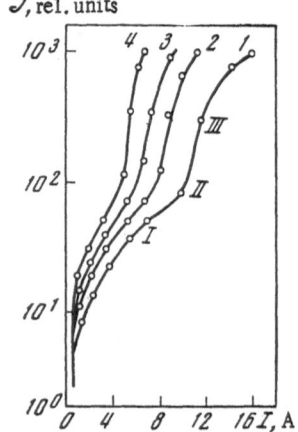

Fig. 23. Influence of a transverse magnetic field on the lux−ampere characteristics of an InAs diode at 77°K. H (kOe): 1) 0; 2) 3; 3) 6; 4) 10.

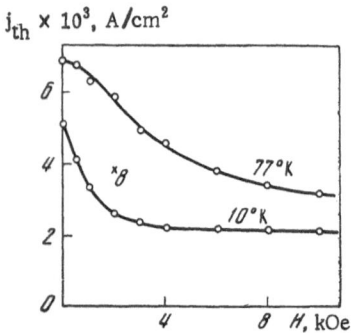

Fig. 24. Dependences of the threshold current density on the transverse magnetic field applied to an InAs diode at 77 and 10°K. The ordinate scale for the curve representing 10°K is magnified by a factor of 8.

stronger fall of the threshold current density with increasing field and saturation was reached at 4 kOe. The dependence of the threshold current density on the transverse magnetic field (the threshold was deduced from the spectra) was not affected by the direction of the field relative to the resonator axis: this direction could be parallel or perpendicular to the axis.

All the observed phenomena could not be explained simply by changes in the energy band structure under the action of magnetic fields because these phenomena were observed only for transverse fields and the energy bands of InAs were known to be isotropic. Moreover, the quantization condition (7) was only weakly satisfied at liquid helium temperatures in fields as high as $H_{max} = 10$ kOe.

The experimental results could be explained by invoking the mechanism of magnetic deflection of injected carriers at right angles to the current [74]. We shall show that one of the results of the action of the Lorentz force on carriers is a reduction in the diffusion length. In a laser diode this reduces the thickness of the active region without altering the number of the injected carriers, i.e., it raises the carrier density. This lowers the threshold current in a transverse magnetic field.

The influence of a transverse magnetic field on injected carriers follows from the equations describing the flow of an electric current in a magnetic field [75, 76]. If the x axis is perpendicular to the p–n junction plane and the field H is directed along the z axis, we find that the Lorentz force deflects the carriers in the direction of the y axis. In regions close to ohmic contacts the Hall emf is small so that the contacts short-circuit the Hall emf, i.e., we have $E_y = 0$. This assumption can be adopted also in our case. If we also postulate that the carrier density does not vary along the junction, i.e., $\partial n / \partial y = \partial p / \partial y = 0$, the equations for the current become

$$\left.\begin{aligned}
j_{nx} &= en\mu_n E_x + eD_n \frac{\partial n}{\partial x} + \mu_n H_z j_{ny}, \\
j_{px} &= ep\mu_p E_x - eD_p \frac{\partial p}{\partial x} - \mu_p H_z j_{py}, \\
j_{ny} &= -\mu_n H_z j_{nx}, \\
j_{py} &= \mu_p H_z j_{px},
\end{aligned}\right\} \tag{24}$$

where μ_n, μ_p, D_n, and D_p are the electron mobilities and the diffusion coefficients, respectively.

Solving the system (24) for j_{nx} and j_{px}, we obtain

$$\left.\begin{aligned}
j_{nx} &= en \frac{\mu_n}{1 + \mu_n^2 H_z^2} E_x + e \frac{D_n}{1 + \mu_n^2 H_z^2} \frac{\partial n}{\partial x}, \\
j_{px} &= ep \frac{\mu_p}{1 + \mu_p^2 H_z^2} E_x - e \frac{D_p}{1 + \mu_p^2 H_z^2} \frac{\partial p}{\partial x}.
\end{aligned}\right\} \tag{25}$$

Hence, we can see that the magnetic field reduces the diffusion lengths of electrons and holes by a factor of $(1 + \mu^2 H_z^2)^{-1}$. If we assume that the carrier lifetime is independent of the magnetic field, we find that the diffusion length L is

$$L = L_0/(1 + \mu^2 H^2)^{1/2}, \tag{26}$$

where L_0 is the diffusion length in $H = 0$. Thus, the application of a transverse magnetic field reduces the diffusion lengths of carriers and, consequently, reduces the thickness of the active region near the p–n junction.

It is known [77-79] that the recombination radiation emitted from InAs and GaAs injection diodes is generated in the p-type region. At carrier densities of $\sim 10^{18}$ cm^{-3} in the original n-type material, this means that the radiative transitions occur from the conduction band (or from a tail of the density of states, depending on the population) to a smeared acceptor band, which overlaps partly the valence band [80, 81]. In this situation narrowing of the active region because of the reduction of L in a transverse magnetic field fills the conduction band, i.e., it shifts upward the quasi-Fermi level of electrons because $m_n \ll m_p$. Consequently, the spontaneous radiation maximum (or the gain maximum) shifts in the direction of shorter wavelengths (compare Figs. 21a and 22a).

This upward shift of the quasi-Fermi level in the conduction band results also in broadening of the emission spectrum (Fig. 20b) and it raises the integrated intensity (Figs. 20a and 22b).

However, it is clear from our figures that all these effects depend on the direction of the transverse magnetic field. The effect is always greater for the fields H$^+$, when carriers are deflected toward the face being investigated. Clearly, this is due to the fact that a thin absorption layer forms at this face when a field H$^-$ is applied because carriers are deflected into the crystal.

The threshold is independent of the direction of the transverse magnetic field because it is governed primarily by the amplification of light along the whole p–n junction. The stronger field dependence of the threshold current density at helium temperatures (Fig. 24) is due to the higher electron mobilities, as indicated by Eq. (26).

The application of a transverse magnetic field also shifts each of the stimulated emission modes of InAs toward shorter wavelengths. Figure 25 shows the energy positions of various modes obtained in transverse fields up to 10 kOe applied at 10°K. The rate of shift is approximately the same for all the modes and it amounts to $(\partial \mathcal{E}/\partial H)_{T, \lambda} = (3 \pm 1) \times 10^{-8}$ eV/Oe. This shift is not observed for longitudinal fields up to 10 kOe (within the limits of the experimental error).

Substituting the above value of $(\partial \mathcal{E}/\partial H)_{T, \lambda}$ into Eq. (21a), we find that the dispersion of the refractive index is $(\partial n/\partial H) = -2.9 \times 10^{-7}$ Oe^{-1} for transverse magnetic fields up to 10 kOe.

An estimate based on the formulas given in [38] shows that $|\partial n/\partial H|_\lambda < 2 \times 10^{-8}$ Oe^{-1} if we allow only for the explicit dependence of n on H as a result of the interaction of light with the electron gas in a semiconductor subjected to a magnetic field.

This difference between the experimental and calculated values of $(dn/dH)_{T, \lambda}$ and the dependence on the direction of the magnetic field can again be explained by invoking the model of magnetic deflection of the injected carriers. As pointed out earlier, this results in compression of the active region and a consequent upward shift of the quasi-Fermi level of elec-

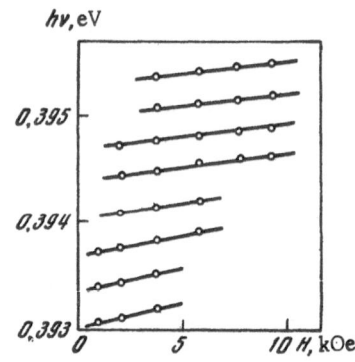

Fig. 25. Dependences of the energies of various InAs laser modes on the transverse magnetic field at 10°K.

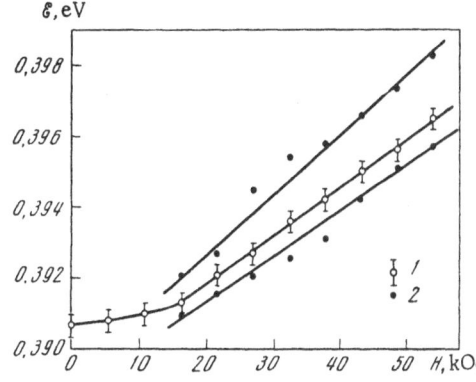

Fig. 26. Dependences of the energies at the maxima of the spontaneous and coherent radiation
spectra of an InAs diode No. 527 on the longitudinal
magnetic field applied at 10°K: 1) spontaneous radiation (j ≈ 280 A/cm²); 2) coherent radiation (j ≈
1.1 × 10³ A/cm²).

trons F_n. This rise of F_n shifts the absorption band edge toward higher energies. Such a
shift reduces the refractive index.

C. Strong Magnetic Fields. The influence of quantizing fields up to 56 kOe on the spontaneous and coherent radiation emitted by InAs diodes was investigated at ~10°K. The measurements were carried out in longitudinal fields in order to avoid the appearance of the mag-
netoresistance in the bulk of a sample.

The influence of a magnetic field on the energy at the maximum of the spontaneous radiation spectrum is demonstrated in Fig. 26. We found that there was a wide range of magnetic fields in which the shift of the maximum was linear. The slope of this linear region was
$(1.35 \pm 0.15) \times 10^{-7}$ eV/Oe. Since the effective mass of holes at the acceptor levels was much
higher than the effective mass of electrons in the conduction band, we could assume that the
slope was governed primarily by the cyclotron mass of electrons. Then, we found that if the
spin splitting was ignored, Eq. (3) led to the following expression for the first Landau level
($l_n = 0$) in the conduction band:

$$\frac{d\mathcal{E}}{dH} \approx \frac{1}{2} \frac{e\hbar}{m_{cn}^* c} = \beta \frac{m_0}{m_{cn}^*} = (1.35 \pm 0.15) \cdot 10^{-7} \text{ eV/Oe}.$$

Hence, we found that the cyclotron effective mass of electrons was $m_{cn}^\bullet = (0.043 \pm 0.005) m_0$,
which was in good agreement (for a carrier density of $\sim 2 \times 10^{18}$ cm^{-3}) with the value deduced
from the Faraday rotation [82] and radiative recombination [83]. This effective mass could
reasonably be attributed to radiative transitions from the first Landau level ($l_n = 0$) in the
conduction band to the acceptor band. The effective mass for the higher Landau levels should
be $(2l + 1)$ times greater.

Extrapolation of the linear region to H = 0 gave a gap of 0.3881 eV, which represented
the separation between the density-of-states tail of the conduction band and the acceptor band.
The difference between the energy of the emitted photons $\hbar\omega = 0.3907$ eV in H = 0 and the extrapolated value was 2 meV and it was due to the filling of the band.

The half-width of the spontaneous radiation line was $\sim 9 k_0 T$ and it was independent of
the magnetic field for a constant current through a diode.

When the current through a diode exceeded the threshold value by a factor of about 2,
a strong magnetic field split the coherent radiation spectrum into two groups of modes with
centers of gravity dependent on the field (Fig. 26). The splitting became noticeable in fields
$\gtrsim 20$ kOe. The intensity of the short-wavelength group was 1.5-2 times higher than the intensity of the long-wavelength group and the former was practically independent of the field.
However, the intensity of the long-wavelength modes increased slightly with the field.

The effect in question was difficult to observe because the coherent radiation emitted by a p–n junction appeared in the form of filaments, as demonstrated clearly in measurements of the angular divergence of the radiation. This inhomogeneity of the emission gave rise to modes (or groups of modes), which were shifted slightly relative to the main emission band.

The splitting in the emission spectra in strong magnetic fields could be explained by the splitting of the lower Landau level in the conduction band ($l_n = 0$) into two spin sublevels.

Then, using the slopes of the lines in Fig. 22, which were 1.2 and 1.7×10^{-7} eV/Oe, and the effective mass $m_{cn} = 0.043 m_0$, we applied Eq. (3) to the determination of the absolute value of the g factor of the conduction electrons and found that $|g_n| \approx 8$. The g factor calculated from Eq. (13) was $g_{theor} \approx -7$ for $m_n = 0.043 m_0$, $\Delta = 0.43$ eV, and $E_g = 0.42$ eV.

This value of the g factor was approximately half the value at the band edge (Table 3). Here, as in the case of the effective mass, the conduction band parameters were affected strongly by heavy doping.

We shall conclude this section with several general remarks.

The results of experiments carried out in magnetic fields were independent of the type of impurity (Zn, Cd) used in the diffusion treatment. The spontaneous emission wavelength did not vary greatly in the temperature range 4–77°K. In this temperature range the photon energy varied, depending on the diode, from 0.389 to 0.392 eV in the case of cadmium-doped samples and from 0.394 to 0.404 eV in the case of zinc-doped samples.

The threshold current density was $(7-8) \times 10^3$ A/cm^2 at 77°K and it fell to 1.6×10^2 A/cm^2 at 4.2°K. The threshold was measured also at 10 and 30°K. The temperature dependence of the threshold current could be represented by $I_{th} \sim \exp(-T/T_0)$, where $T_0 \approx 20$°K. This exponential temperature dependence of the threshold current demonstrated the great importance of the density-of-states tails in the radiative transitions.

The dependence of the laser emission frequency on the magnetic field could be used to tune magnetically the emission frequency of infrared radiation sources.

Continuous variation of the emission frequencies of individual modes (Fig. 25) in weak transverse magnetic fields was interesting: this variation could reach ∼ 50 GHz for emission lines ≲ 20 GHz wide (the width was limited by the resolution of the instruments employed). In quantizing fields this tuning range was greater and it amounted to 3.11–3.17 μ (Fig. 26). However, in this case the tuning was discrete and the changes were equal to the mode separation ∼ 25 Å (some modes appeared and other disappeared).

The dimensions of the emitting region were estimated by investigating the spatial distribution of the radiation intensity in the far zone. Figure 27 shows the angular variation of

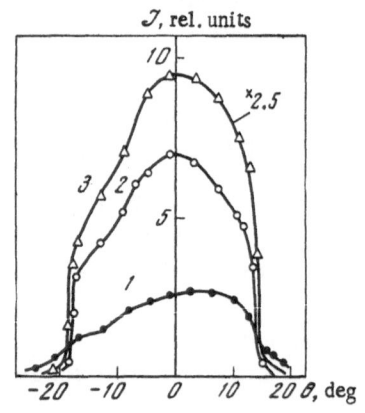

Fig. 27. Distribution of the intensity of the coherent radiation emitted by an InAs diode in a plane perpendicular to the p – n junction, plotted for different currents through the diode I (A): 1) 0.5; 2) 3; 3) 10. Diode No. 5, T = 4.2°K.

the intensity of the coherent radiation in a plane perpendicular to the junction: the angular width was found to be $\theta \approx 30°$ and it depended weakly on the current. This corresponded to an emitting region $\sim 6\ \mu$ wide. In the p–n junction plane there were usually several beams whose existence indicated an inhomogeneity of the laser action in the p–n junction. In this case the divergence was 6°, which corresponded to emission from a small part of the p–n junction ($\sim 30\ \mu$).

§2. Radiation Emitted from InSb Injection Lasers in Strong Magnetic Fields. Position of Light-Emission Region

The magnetic-field dependence of the coherent emission frequency of diffused p–n junctions in indium antimonide was investigated in [15, 84]. The laser action was obtained only in longitudinal magnetic fields exceeding 20 kOe. The method of double injection of carriers into pure indium antimonide crystals [85–87] made it possible to reduce the minimum longitudinal magnetic field necessary for the laser action to $H_{min} = 3.5$ kOe [88]. It was established in [89, 90] that in this case the laser action occurred in a degenerate electron–hole plasma and the value of H_{min} was limited by the pinch effect, i.e., by the compression of plasma by its own magnetic field [91].

The magnetic compression of an injected plasma is usually investigated by measuring the plasma conductivity (current–voltage characteristic), determining the recombination radiation spectra of the plasma, and observing instabilities in the plasma conductivity due to the pinch effect.

We studied the pinch effect by the method of scanning the recombination radiation of an electron–hole plasma in a crystal [92] and found [93] that thin ($\sim 100\ \mu$) crystals emitted light uniformly. This was due to the absence of the pinch effect and it was in agreement with the results reported in [94], where the laser action was first observed in indium antimonide in the absence of a magnetic field. Employing thin crystals, we were able to tune the coherent emission wavelength of an electron–hole plasma in InSb within the range 5.02–5.30 μ by varying a longitudinal magnetic field from zero to 54 kOe [16].

An electron-hole plasma was generated by the double injection of carriers into pure indium antimonide crystals formed into p^+–i–n^+ structures. The distance between the contacts was varied from 0.1 to 0.5 mm. The original crystals usually had p-type conduction with a hole density of $\sim 10^{13}$ cm^{-3} and a mobility $\mu \approx 6 \times 10^3$ cm$^2 \cdot$ V$^{-1} \cdot$ sec^{-1} at 77°K. Carriers were injected in the form of rectangular current pulses of 10^{-5}–10^{-6} sec duration. The measurements were carried out at liquid helium temperature. The forward branch of the current-voltage characteristic was S-shaped in the low-current region, which indicated the occurrence of double injection throughout a crystal at high currents.

In the current range 2–15 A the injected carrier density in crystals of $\sim 2 \times 10^{-3}$ cm^2 cross section was n = $j/ev_{dr} \approx (0.6-5) \times 10^{16}$ cm^{-3}, where the drift velocity was assumed to be $v_{dr} \approx 10^6$ cm/sec. At these carrier densities and for an electron temperature T = 11–18°K (there was some heating of the electron gas [93]), the electron gas was strongly degenerate: $F/k_0 T_e = 8$–18.

A degenerate plasma is obtained if the Fermi–Thomas screening radius is greater than the distance between neighboring charged particles, i.e., if the kinetic energy of the particles F is greater than the potential energy due to the Coulomb interaction. This means that the number of particles N in a volume bounded by the screening radius should be greater than

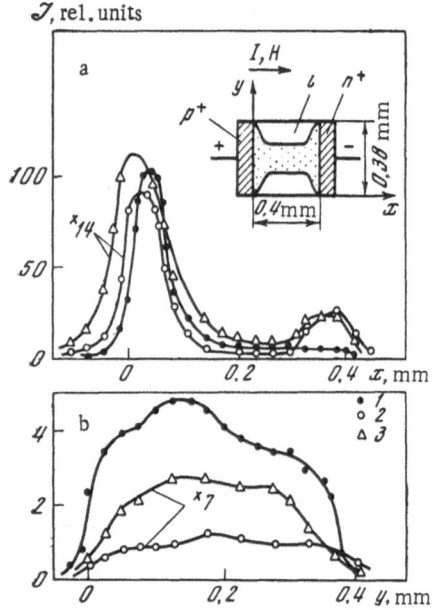

Fig. 28. Distribution of the emission from an InSb crystal 0.4 mm thick along the direction of the current (a) and at right angles to the current (b). The curves demonstrate the influence of a longitudinal magnetic field: 1) I = 3.4 A, H = 0; 2) I = 12 A, H = 0; 3) I = 12 A, H = 410 Oe. The diode area is S = 1.6 × 10⁻³ cm².

unity. In the case of a degenerate plasma, we have [95]

$$N = 3.76 \times 10^{-13} \left(\frac{\varepsilon_0 m_0}{m^*} \right)^{3/2} n^{1/2}. \tag{27}$$

For the injected-carrier densities considered above, we find that N ≈ 2.

A. Position of Light-Emission Region. Influence of Weak Magnetic Fields on Radiation Intensity. Figure 28 shows the distribution of the intensity of the recombination radiation emitted by an InSb crystal 0.4 mm thick along (a) and at right angles (b) to the direction of current. The inset shows schematically a section of a diode and a possible spatial distribution of plasma for $I > I_{cr}$. In both cases the scanning was concentrated in the middle part of the crystal. The current through a diode was either lower or higher (represented in Fig. 28 by black dots and open circles, respectively) than the critical pinch current $I_{cr} \approx 11$ A. The pinch current was deduced from the current–voltage characteristic but the onset of the pinch effect was observed at currents approximately half the value of I_{cr} [93]. The triangles in Fig. 28 represent the distribution of the recombination radiation obtained for a current $I > I_{cr}$ in a longitudinal magnetic field H = 410 Oe.

It is clear from Fig. 28a that for $I < I_{cr}$ the radiation intensity was higher than in other cases and the emission was observed throughout a crystal 0.4 mm long. A bright emission region (~ 70 μ long) was observed at the anode and the intensity in this region was about 20 times higher than elsewhere in the crystal. In the $I > I_{cr}$ range the radiation intensity fell by a factor exceeding 14 and it was concentrated mainly at the contacts. In this case the application of an external magnetic field increased strongly the radiation intensity, particularly near the center of the crystal.

Scanning of the radiation emitted at right angles to the current (Fig. 28b) showed that in the $I < I_{cr}$ range the emission was distributed fairly uniformly across a crystal but in the $I > I_{cr}$ case the radiation intensity fell considerably and the emission region was somewhat narrower. The application of a longitudinal magnetic field in the $I > I_{cr}$ case raised the radiation intensity by a factor of about 2.

A study of the distribution of the radiation intensity along the current indicated that when the thickness was reduced, the region of the dip in the curve for $I > I_{cr}$ became narrower and

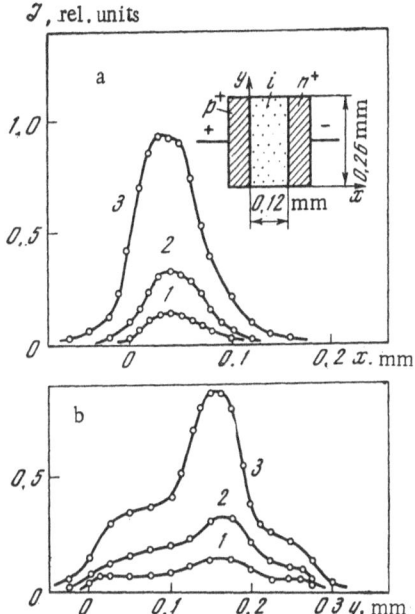

Fig. 29. Distribution of the emission from an InSb crystal 0.12 mm thick along the direction of the current (a) and at right angles to the current (b). I(A): 1) 1.5; 2) 3; 3) 10. The area of the diode is $S = 1.3 \times 10^{-3}$ cm^2.

it disappeared completely when the thickness was reduced to ~ 0.1 mm (Fig. 29). The inset shows schematically a section through a thin diode.

The observed intensity distribution was due to the pinch effect in the electron–hole plasma, which concentrated carriers in a thick crystal into a plasma filament. Since carriers were injected uniformly throughout the contact area, the compression of the plasma into a filament was impossible in the regions near the contacts. The size of this contact region was governed by the distance which was traversed by drifting carriers during the formation of a filament, $l \sim v_{dr} \tau \approx 10^{-2}$ cm (it was assumed that the filament formation time was $\tau \approx 10^{-8}$ sec). Thus, the distribution of carriers over the cross section of a sample became of the dumbbell shape (shown in the inset in Fig. 28a). Since the radius of a plasma filament was $\sim 10^{-2}$ cm [91], the carrier-depleted i-type region was of the same size ($\sim 1.5 \times 10^{-2}$ cm). Consequently, this region reduced strongly the intensity of the radiation emitted from the plasma filament although the compression of the carrier flux increased this intensity somewhat (the compression increased the density of states at the Fermi level because of the rise in the carrier density).

A longitudinal magnetic field destroyed the pinch effect giving rise to a pressure which was additional to the internal gas–kinetic pressure in the plasma. This was why the intensity increased somewhat in a field of 410 Oe.

The influence of a longitudinal magnetic field on the integrated (over the surface) intensity of the radiation emitted from the same crystal is shown in Fig. 30a. Clearly, in the range $I > I_{cr}$ the intensity increased in fields up to ~ 1 kOe. In stronger fields the intensity varied only slowly differing slightly in value for different relative orientations of the current and magnetic field. In the case of InSb the range 1–4 kOe represented the transition to the quantum region and we shall not discuss it here. The increase in the radiation intensity in fields up to 1 kOe was due to a gradual expansion of the plasma filament until it reached the walls of the crystal. The plasma then filled the whole crystal. It should be noted that the destruction of the pinch, deduced from the current–voltage characteristics of a diode in a longitudinal magnetic field [91], also occurred in fields of ~ 1 kOe.

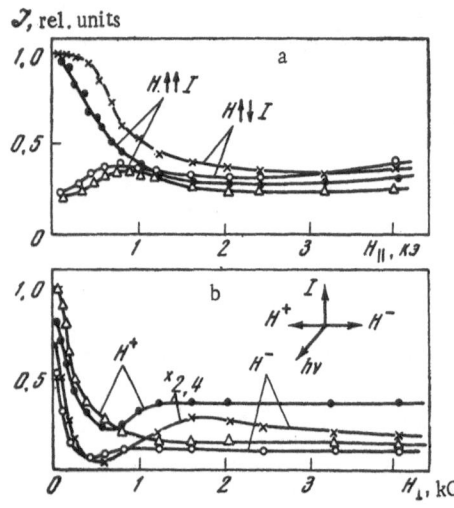

Fig. 30. Dependences of the intensity of the integrated radiation emitted from an InSb crystal 0.4 mm thick on the longitudinal (a) and transverse (b) magnetic field. The black dots and crosses correspond to I = 3.4 A and the open circles and triangles correspond to I = 12 A.

In the $I < I_{cr}$ range the intensity of the radiation emitted in a longitudinal field fell up to ~ 2 kOe and the actual effect of the field depended on the relative directions of the field and current. One of the possible explanations of this effect is the diffusion recombination in weak magnetic fields considered in [96]. According to this mechanism, the large difference between the effective masses of electrons and holes in indium antimonide ($m_p / m_n \approx 46$) results in a basically inhomogeneous spatial distribution of charges in an electron–hole plasma. Then, in the zeroth approximation, the low-mobility holes can be regarded as electron-recombination centers. In weak magnetic fields the rate of recombination slows down and the radiation intensity $\mathcal{J}$ becomes

$$\mathcal{J} \propto \frac{\mu}{\varepsilon_0} \cdot \frac{1 + \frac{1}{3}\left(\frac{\mu H}{c}\right)^2}{1 + \left(\frac{\mu H}{c}\right)^2} \, np, \tag{28}$$

where n and p are, respectively, the densities of the injected electrons and holes; ε_0 is the permittivity of the investigated crystal. This behavior of the intensity is in qualitative agreement with the experimental results. However, this model fails to explain the dependence of the radiation intensity on the relative directions of the current and field.

The stronger dependence of the radiation intensity on a transverse magnetic field (Fig. 30b), observed for $I < I_{cr}$ and for $I > I_{cr}$, was evidently due to the heating of carriers. This was deduced from the emission spectra obtained in fields of ~ 50 Oe [91]. The heating in the transverse field lifted the degeneracy and destroyed the plasma filament.

B. Emission of Coherent Radiation in Strong Magnetic Fields. We mentioned earlier that diodes ~ 0.1 mm thick emitted radiation from the whole of their area and could produce coherent radiation in the absence of an external magnetic field [94]. Therefore, it seemed interesting to study the coherent emission spectra of such diodes in a wide range of magnetic fields from zero to several tens of kilo-oersteds in order to determine some of the energy structure parameters of indium antimonide during injection of nonequilibrium carriers into pure crystals and to find whether it would be possible to tune the stimulated emission frequency by the application of a magnetic field. Indium antimonide is an attractive material for frequency tuning experiments because the effective mass of electrons is low and this should ensure a large shift of levels in an external magnetic field.

The investigation was carried out in longitudinal magnetic fields up to ~ 50 kOe, which were generated by the superconducting solenoid described in Chap. II, § 3. The use of this

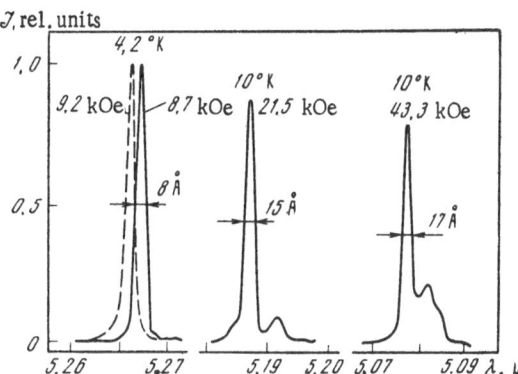

Fig. 31. Coherent emission spectra of an
InSb laser obtained in different magnetic
fields under conditions close to single-
mode emission. The current density at
4.2°K was j = 570 A/cm^2 and at 10°K it was
j = 1.2 × 10^3 A/cm^2.

solenoid was particularly convenient in the case of indium antimonide lasers because these
lasers operated only at helium temperature, which conveniently matched the conditions necessary for the operation of the solenoid.

Figure 31 shows the coherent emission spectra of an indium antimonide laser recorded
in magnetic fields of different intensities at two temperatures 4.2 and 10°K. In these experiments the injection level was slightly above the threshold value so that single-mode (or nearly
single-mode) emission was possible. An increase of the magnetic field generated modes with
shorter wavelengths. The thermal shift at helium temperatures was much weaker than at
higher temperatures and it had the opposite direction. Moreover, when the magnetic field was
altered slightly, the individual modes also shifted in the direction of shorter wavelengths (see
the spectrum at 4.2°K). The widths of the coherent emission lines representing single modes
were governed by the transfer function of the spectrometer.

The relative radiation intensity fell somewhat with rising magnetic field (Fig. 31). This
was probably due to the heating of carriers because the resistance of the laser crystal increased in transverse magnetic fields. The presence of a small transverse component of the
field was indicated by a small (several percent) increase of the voltage drop across the crystal when a field of 50 kOe was applied. The fall of the intensity for a fixed value of the current through the diode could also occur because of a downward shift of the Fermi level F_n
and because of a partial lifting of the degeneracy. For example, when the injected carrier
density was about 8 × 10^{15} cm^{-3} and the crystal was subjected to H = 50 kOe, Eq. (10) predicted that $F_n \approx 1$ meV, which was only slightly higher than the thermal energy of carriers
$k_0 T$ at T $\approx$ 10°K.

When the magnetic field was increased from zero to 54 kOe, the emission wavelength
varied quasicontinuously (to within the spacing between neighboring Fabry–Perot resonator
modes) between 5.30 and 5.02 μ, which corresponded to a variation of the photon energy ℏω
from 234 to 247 meV. The dependence of the energy of the dominant laser mode on the magnetic field is plotted in Fig. 32. The abscissa gives the magnetic fields at which the mode

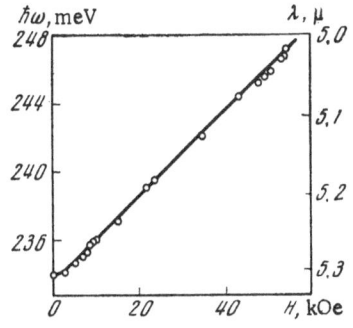

Fig. 32. Dependence of the energy of the dominant
mode emitted by an InSb laser on the magnetic field
intensity.

intensity was highest. Thus, the mode energy corresponded to the maximum of the gain profile. We assumed that this maximum coincided with the peak of the recombination radiation spectrum of indium antimonide. It is clear from Fig. 32 that in fields H $\gtrsim$ 4 kOe the plotted dependence was linear. Bearing in mind that at helium temperatures fields of several kilo-oersteds were already quantizing for the conduction band of indium antimonide, we could explain the observed radiation by the interband recombination of electron–hole pairs between the s = +1/2 spin sublevel of the lowest Landau level in the conduction band and the valence band [88]. The experimental value of the rate of shift of the emission line was 0.25 meV/kOe. The substitution of the injected-electron density $\sim 7 \times 10^{15}$ cm^{-3} (for j = 1.2 $\times$ 10^3 A/cm^2) and of the electron g factor for this density g $\approx$ -48 (see [93]) into Eq. (3) yielded the effective electron mass $m_n \approx 0.015 m_0$. This was a reasonable value because the nonparabolicity of the conduction band began to be manifested at the injection levels employed.

When the dependence $\hbar\omega = f(H)$ was extrapolated to H = 0 (Fig. 32), it was found that the forbidden-band width for this electron density amounted to E_g = 233.5 meV, i.e., this width was 2.2 meV smaller than for pure crystals [98]. The slight reduction in the forbidden-band width could be due to the Coulomb interaction between carriers [99].

The relative change in the emission frequency of the coherent radiation was $\sim$5%. This change was the highest value likely to be obtained for well-known semiconductor lasers because indium antimonide had the lowest effective carrier mass. In principle, this narrow tuning range could be approximately doubled by utilizing the upper (s = -1/2) spin sublevel of the lowest Landau level in the conduction band. However, this would require much higher densities of the injected carriers. According to Eq. (10a), the required density would be $\gtrsim$ 5 $\times$ 10^{16} cm^{-3} for H = 50 kOe.

The shift of a single mode in a magnetic field was an interesting point in connection with the attempt to achieve continuous frequency tuning (Fig. 31). This shift was 15–18 Å/kOe, which was in agreement with the results reported in [15]. In view of the fact that the mode separation was $\Delta\lambda \approx$ 42 Å and the length of a crystal was l = 0.59 mm, the change in the refractive index in a magnetic field H = 10 kOe calculated using Eq. (21a) was $(\partial n/\partial H)_\lambda = -1.7 \times 10^{-6}$ Oe^{-1}.

The effective refractive index was n$^\bullet$ = n - λ $(\partial n/\partial\lambda)$ = 5.2-5.6 and it depended on the duration of the current pulses, because of the heating of the crystal.

§ 3. Spontaneous and Coherent Radiation Emitted from

p—n Junctions in PbSe

Lead chalcogenides are more suitable materials for infrared lasers than compounds such as InSb. This is due to the following factors. First of all, the density of states in the conduction and valence bands of lead salts is higher because of the larger effective mass and because of the presence of four valleys at the points L; therefore, the interband optical tansitions are more likely and the radiative recombination is relatively strong. Secondly, the impact interband recombination processes in lead chalcogenides are not highly efficient because the effective masses of electrons and holes are approximately equal and the permittivity is high. The interband impact recombination does not play any significant role in lead chalcogenides, even at the high injection levels encountered in lasers. Thirdly, the loss of photons due to the absorption by free carriers is relatively small because of the high carrier mobility.

An investigation was made of p—n junctions prepared from n- and p-type PbSe with carrier densities (1.2-5.0) $\times$ 10^{18} cm^{-3} and mobilities (1-3) $\times$ 10^4 cm$^2 \cdot$ V$^{-1} \cdot$ sec^{-1} at 77°K. These p—n junctions were prepared by the diffusion of selenium or lead from a PbSe powder in sealed quartz ampoules. Resonators were formed by cleaving along the (100) planes. The emission

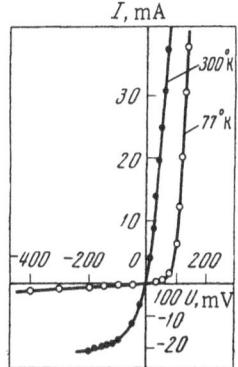

Fig. 33. Current—voltage characteristics of a PbSe diode recorded at temperatures of 77 and 300°K.

of spontaneous radiation was investigated in the case of diodes which had high laser thresholds. The measurements were carried out in a longitudinal magnetic field at temperatures of 4.2 and 10°K. These diodes were excited with current pulses of 1–10 μsec duration and ~ 100 Hz repetition frequency.

Typical current—voltage characteristics obtained at 77 and 300°K are shown in Fig. 33. The dependence of the forward current on the voltage was given by $I = I_0 \exp(eU / rk_0 T)$, where r varied from 2.4 to 3. This indicated that a considerable proportion of the injected carriers recombined in the p–n junction depletion region.

A. Spontaneous Radiation. Figure 34 shows the positions of the recombination radiation maxima as a function of the applied magnetic field up to 50 kOe. At ~ 10°K there was usually one spontaneous emission line L_1, which shifted linearly toward high energies at a rate of $+1.0 \times 10^{-7}$ eV/Oe. This rate of shift was in agreement with the results reported in [10, 100]. At 4.2°K two lines L_1 and L_2 were observed [101]. The line L_2 shifted more rapidly with the magnetic field: its shift was approximately linear with a slope of $+2.8 \times 10^{-7}$ eV/Oe.

In order to identify the transitions responsible for the lines L_1 and L_2, we determined their polarization in a field $H \approx 10$ kOe, when the intensities of the two lines were approximately equal. It was found that the line L_1 was linearly polarized parallel to the magnetic field, whereas the line L_2 was polarized at right angles to the field. Applying selection rules similar to those which govern the polarization of light in the Zeeman effect in atoms when light is traveling at right angles to the magnetic field (Fig. 10), we found that the line L_1 should correspond to electron transitions from the valence to the conduction band in which the spin was conserved, whereas the line L_2 was due to similar transitions but accompanied by spin flip.

The absence of splitting of the line L_2 in the magnetoabsorption experiments was attributed in [42] to approximately equal and opposite g factors of the conduction and valence

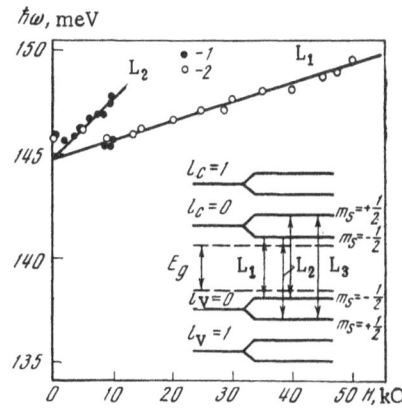

Fig. 34. Positions of the maxima in the recombination radiation spectrum of PbSe plotted as a function of a magnetic field applied along the [100] axis at T = 4.2°K (1) and 10°K (2). The lower part of the figure shows the energy spectrum of PbSe in a magnetic field.

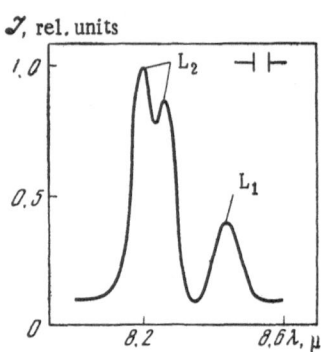

Fig. 35. Spectrum of the recombination radiation emitted by a p – n junction in PbSe subjected to H = 9.2 kOe at T = 4.2°K. The duration of the current pulses was $\tau = 10$ μsec and their repetition frequency was 120 Hz; L_1 denotes a transition in which spin is conserved and L_2 denotes a transition accompanied by spin flip.

bands. We carried out the first detailed study of the transition L_2 at 4.2°K. We found that the line L_2 was split quite strongly (~ 0.5 meV in a field of 9.2 kOe) into two identically polarized peaks (Fig. 35). Since the splitting exceeded only slightly the resolution limit of the apparatus, we could only draw an important qualitative conclusion that the line L_2 was due to two transitions of almost equal energies and these transitions corresponded to spin flip. Thus, the results of the polarization measurements demonstrated convincingly that the g factors of the valence and conduction bands of PbSe had opposite signs. The difference between the absolute values of the g factors gave rise to the experimentally observed splitting of the line L_2.

These conclusions on the nature of the observed transitions lead us to postulate the possible energy spectrum of PbSe in quantizing magnetic fields, shown as an inset in Fig. 34. In this figure the g factor of the conduction band is taken to be positive and that of the valence band is assumed to be negative. The arrows show possible transitions between the spin-split Landau levels ($l_n = l_p = 0$) in the valence and conduction bands. We observed only the L_1 and L_2 transitions.

According to Eq. (9), the energy of photons emitted in a magnetic field as a result of transitions involving the $l_n = l_p = 0$ Landau levels can be expressed in the form

$$\hbar\omega = E_g + \left[\frac{m_0}{m_r} \pm (g_n \pm g_p)/2 \right]\beta H,\qquad(29)$$

where m_r is the reduced effective mass given by the expression

$$\frac{1}{m_r} = \frac{1}{m_n} + \frac{1}{m_p};\qquad(30)$$

m_n, m_p, g_n, and g_p are the effective masses and the g factors of the conduction and valence bands, respectively; E_g is the forbidden band width; β is the Bohr magneton; H is the magnetic field. The slope of the line L_2 can be used directly to find the reduced effective mass $m_r = 0.021 m_0$. Then, using this value and the slope of the line L_1, we can find the g factors: $|g_n| = |g_p| = 30$, which are obtained ignoring the splitting of the line L_2. It should be pointed out that the values of the effective masses and g factors calculated in this way differ from the corresponding values obtained along the [100] direction in magnetoabsorption experiments [42]: $m_r = 0.029 m_0$ and $|g_n| = 22$. This difference is quite likely because in such experiments the high populations of the bands mean that the Landau levels of number 1 or higher may be manifested. Therefore, away from the band edges the values of m_r are higher and the g factors are lower because of the band nonparabolicity.

Extrapolation of the lines L_1 and L_2 to H = 0 gives the forbidden band width of PbSe, which is $E_g = 0.145$ eV. This value of E_g agrees with the value obtained in [10, 100] by a similar method.

It is known that in the case of a narrow forbidden band and a strong spin—orbit interaction both the g factor and the effective mass may be governed primarily by the interaction between the highest valence band and the lowest conduction band. According to the two—band model, the g factor is considerably greater than 2 and, if the principal extrema of the conduction and valence bands are only spin-degenerate, the g factor is related to the cyclotron effective mass m by the simple expression (14). It is important to note that in contrast to the Kane model of InSb, we are assuming here that there is no degeneracy (apart from the spin) of the conduction and valence band edges. It follows from Eq. (14) that the spin splitting of the Landau levels is equal to the separation between these levels. Consequently, an analysis of the energy spectrum in accordance with the two—band model [35] shows that the position of the lowest spin Landau sublevel in the conduction band of PbSe ($l_n = 0$, $s = -1/2$) and of the highest Landau sublevel in the valence band ($l_p = 0$, $s = -1/2$) should not be affected by the magnetic field. However, the other sublevels should exhibit degeneracy: $\mathcal{E}(l, s = +1/2) = \mathcal{E}(l + 1, s = -1/2)$. This means that the energy of the transition L_1 should be independent of H. The observed relatively slight shift of the line L_1 in a magnetic field shows that the two—band model does not provide a completely correct description [102].

The observed shift can be explained bearing in mind that actually there are six closely spaced bands near the forbidden gap at the point L in the Brillouin zone (Fig. 3), and the interaction between these bands reduces the g factor compared with the case when only two bands interact. Calculations of the energy structure of PbSe have shown [40, 103, 104] that the other four bands at the point L are separated by about 1.5-2 eV from the main conduction and valence bands.

Allowance for the influence of other bands in the calculation of the magnetic levels in the conduction band of bismuth is made in [105]. It is shown that the degeneracy of the levels is lifted and that the levels forming the minimum energy gap become dependent on the magnetic field (Fig. 36). The analysis developed in [102, 105] for bismuth is based on assumptions which are equally applicable to lead chalcogenides.

We shall now consider the influence of an external magnetic field on the intensities of the L_1 and L_2 lines in the case when the current through a diode was kept constant at a value known to be below the laser threshold in the applied field. When the field was increased from $H \approx 5$ kOe to $H \approx 9$ kOe, only the line L_2 was observed and its intensity increased severalfold in this range. The line L_1 appeared near 9 kOe and this intensity rose rapidly with the mag-

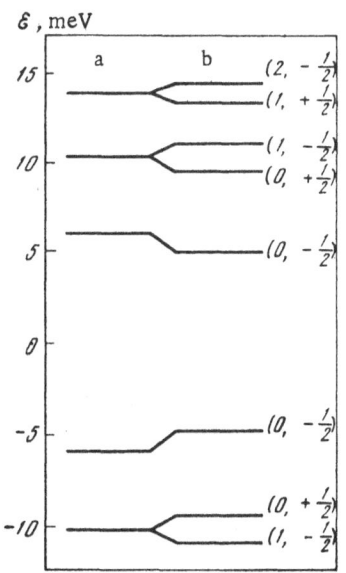

Fig. 36. Calculated spectrum of energy levels of bismuth in a magnetic field [105]: a) two—band models; b) model allowing for the influence of other nearby bands.

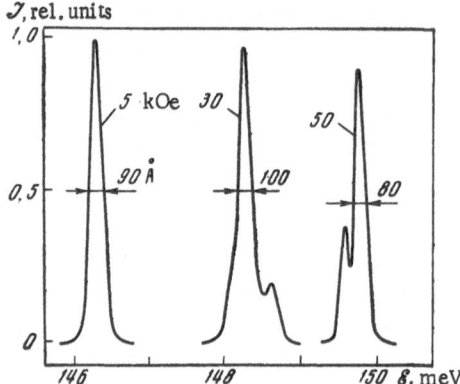

Fig. 37. Coherent emission spectra of a PbSe laser obtained in three different magnetic fields under conditions close to single-mode emission; T = 10°K.

netic field, becoming equal to the intensity of the line L_2 at H $\approx$ 10 kOe. In this range the intensity of the line L_2 increased only slightly. The behavior of the line L_2 was not investigated in the range H > 10 kOe. In fields above 50 kOe at T $\approx$ 10°K the line L_1 disappeared when the current was below the threshold value.

This behavior can be explained if we allow for the magnetic-field dependence of the Fermi level. Estimates based on Eqs. (5), (10), and (10a) show that for a carrier density $\sim 7 \times 10^{15}$ cm^{-3} the Fermi level in a field H $\approx$ 5 kOe is located slightly below the second Landau level. When the field is increased, the Fermi level descends and the density of states at the first Landau level rises so that the intensity of the line L_2 increases and the line L_1 appears and its intensity also rises with the field. According to our estimates the line L_2 should be observed at T = 4.2°K in fields up to 20 kOe.

B. Coherent Radiation. The coherent emission was investigated using diodes with a threshold current density of 600 A/cm^2 at 4.2°K. The threshold decreased in longitudinal magnetic fields (the reduction was 40% in a field H = 10 kOe). This was due to an increase of the density of states at the Landau levels. At liquid helium temperature the laser action was observed in both transitions L_1 and L_2 whereas at 10°K it was concentrated mainly in L_1. The dependences of the energies of the dominant laser modes on the magnetic field were linear for both transitions and the slopes were the same as in the spontaneous emission case (compare with Fig. 34). When the magnetic field was increased from zero to 50 kOe the coherent emission frequency varied quasicontinuously within a narrow range from 8.3 to 8.5 μ. The physical causes responsible for the narrowness of the tuning range are discussed in subsection A. When the excess of the current over the threshold value was small, the laser operated in or close to the single-model regime (Fig. 37).

A small change in the magnetic field shifted the laser modes in the direction shorter wavelengths (Fig. 38). The observed changes in the emission frequencies of two separate modes

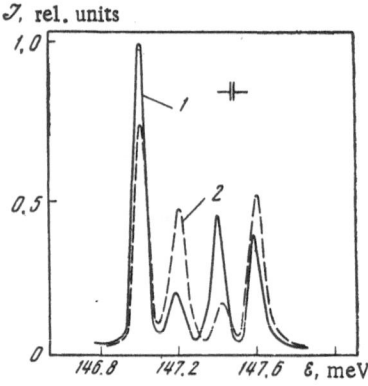

Fig. 38. Coherent emission spectra of a PbSe laser obtained for two similar values of the magnetic field: 1) H = 28.5 kOe; 2) H = 29.3 kOe.

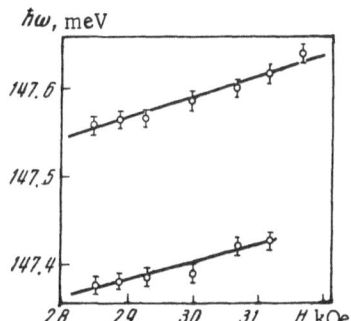

Fig. 39. Frequency shifts of two PbSe laser modes
in the applied magnetic field.

are plotted in Fig. 39. The slope of the lines in this figure is $(\partial\mathcal{E}/\partial H)_{\lambda[100]} \approx 2.3 \cdot 10^{-8}$ eV/Oe. Substituting this value in Eq. (21a), we calculated the dispersion of the refractive index of the active medium of PbSe in magnetic fields. For H = 30 kOe along the [100] direction we obtained $(\partial n/\partial H)_{\mathcal{E}[100]} \approx -1.1 \cdot 10^{-6}$ Oe^{-1}. The effective refractive index n· varied from diode to diode in the range 6.6-7.1.

Polarization measurements of the coherent radiation emitted from PbSe diodes indicated that the modes were not all polarized in the same way but the preferred polarization was in a plane perpendicular to the p—n junction (Fig. 40), where θ is the angle between the p—n junction plane and the plane of polarization of the laser radiation.

CHAPTER IV

MAGNETICALLY TUNED STIMULATED RAMAN EMISSION FROM INDIUM ANTIMONIDE

§ 1. Raman Scattering of Light by Plasmons and Landau Levels in Semiconductors

A. Scattering Cross Section. The appearance of high-power infrared lasers has stimulated further investigations of the scattering of light by free carriers in semiconductors. The inelastic scattering of light was predicted theoretically for various collective [106, 107] and one-particle excitation [108, 109]. In discussing and comparing various types of Raman scattering of light in semiconductors we shall concentrate on the Raman scattering involving optical plasmons and electron transitions between Landau levels in semiconductors [110].

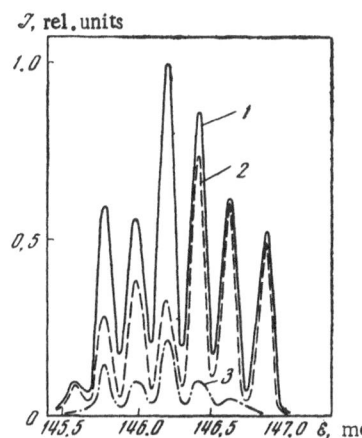

Fig. 40. Polarization of the radiation emitted by
a PbSe laser: 1) without a polarizer; 2) $\theta = 90°$;
3) $\theta = 0$.

When light is scattered inelastically in a crystal, the laws of conservation of energy and momentum are as follows:

$$\hbar\omega_s = \hbar\omega_0 \pm \hbar\omega_e, \tag{31}$$

$$\mathbf{k}_s = \mathbf{k}_0 \pm \mathbf{k}_e, \tag{32}$$

where ω_s, ω_0 and $\mathbf{k}_s$, $\mathbf{k}_0$ are the frequencies and wave vectors of the scattered and incident light, respectively; $\hbar\omega_e$ is the energy of an elementary excitation responsible for the scattering of light and $\mathbf{k}_e$ is the corresponding wave vector. The minus sign corresponds to the Stokes component of the Raman scattering and the plus sign to the anti-Stokes component.

The cross section for the scattering of light by "classical" free electrons in the optical frequency range, where the exchange of energy between photons and electrons can be ignored, is given by the formula (Thomson cross section)

$$\sigma_T = \frac{8\pi}{3}\left(\frac{e^2}{m_0 c^2}\right)^2 = 6.65 \cdot 10^{-25} \text{ cm}^2. \tag{33}$$

The problem of calculation of the scattering of light by free carriers in crystals is much more complex, particularly in the case of narrow-gap semiconductors [111]. This is due to the fact that the velocity of an electron in a crystal is frequently a nonlinear function of its momentum. Therefore, electrons may participate in nonlinear optical processes such as the Raman scattering, which have no analogs in the case of classical electrons. The matrix element for the scattering of light then contains second-order virtual interband terms $-(e/c)(\mathbf{v} \cdot \mathbf{A})$, where $\mathbf{v}$ is the electron velocity operator and $\mathbf{A}$ is the vector potential of the light-wave field. Calculations show that in the range $\hbar\omega_0 < E_g$ the Thomson (elastic) and Raman scattering cross sections of free carriers in a crystal exceed the value of σ_T for free carriers in space, and on approach of $\hbar\omega_0$ to E_g the cross sections of carriers in a crystal increase resonantly.

When a magnetic field is applied, there is also a finite probability of the Raman scattering accompanied by the transfer of an electron from one quantum Landau level to another and by a change in the number of the quantum level. It must be stressed that such a scattering does not occur in the dipole approximation for classically free electrons since the motion of such electrons in a magnetic field is harmonic and the harmonic oscillator model does not predict the Raman scattering. However, the band nonparabolicity of semiconductors makes the motion anharmonic and, consequently, it gives rise to the Raman scattering.

The scattering accompanied by spin flip in a magnetic field can be represented as follows [112, 113]. If the Fermi level of electrons is located between two spin Landau sublevels, light excites an electron from the valence band to the upper spin sublevel and this creates a hole in the valence band (virtual transition $\hbar\omega_0 < E_g$). This is followed by the transfer of an electron from the lower spin sublevel to the valence band. The scattering event is thus completed in accordance with the expressions (31) and (32).

The cross section for the Raman scattering of light by optical plasmons is [114-116]

$$\sigma_{pl} \approx (e^2/mc^2)^2 (\mathbf{k}_p/\lambda_{T\text{-}F})^2 (\hbar\omega_p/F) \text{ cm}^2/\text{sr}, \tag{34}$$

where $\mathbf{k}_p$ is the wave vector of a plasmon; ω_p is the plasmon frequency; $\lambda_{T\text{-}F}$ is the Thomas-Fermi screening radius. For typical electron densities of $\sim 10^{16}$-10^{17} cm^{-3} in InAs the scattering cross section per one electron is $\sim 10^{-24}$ cm^2/sr and the plasmon frequency varies from ~ 70 to 140 cm^{-1}. Consequently, optical plasmons are of interest from the point of view of achieving tunable Raman scattering. Spontaneous Raman scattering by optical plasmons and coupled plasmon–phonon modes has been observed in GaAs [114-116]. The scattering by plasmons has also been found in InAs [117]. However, tuning of the plasmon frequency requires a change in the carrier density in a sample so that we must have either a batch of samples or the carrier density must be varied by injection, which is not always possible. A more convenient method for the plasma frequency tuning is the coupling of plasmon and cyclotron modes in a magnetic field, which gives rise to hybrid resonances in a magnetized plasma. Tunable spontaneous Raman scattering by a magnetized plasma has been reported for GaAs [118].

A much more promising method for achieving tunable Raman scattering is the use of one-particle electron transfers between Landau levels within one band. The spontaneous Raman scattering by such excitations was predicted theoretically in [108, 109] and observed in n-type InSb [119]. Three different processes were observed: 1) in one process the quantum number of the Landu level l changed by 2, i.e., $\Delta l = 2$ (predicted in [108]); 2) in another process l changed by 1, i.e., $\Delta l = 1$ (not predicted theoretically); 3) in the third case the spin number s changed by 1 for the same Landau level, i.e., $\Delta l = 0$, $\Delta s = 1$ (predicted in [109]). In these three cases the frequencies of the scattered light are

$$\Delta l = 2; \quad \Delta s = 0; \quad \omega_s = \omega_0 \pm 2\omega_c; \tag{35}$$

$$\Delta l = 1; \quad \Delta s = 0; \quad \omega_s = \omega_0 \pm \omega_c; \tag{36}$$

$$\Delta l = 0; \quad \Delta s = 1; \quad \omega_s = \omega_0 \pm g\beta H/\hbar. \tag{37}$$

In order to estimate the order of magnitude of the Raman scattering cross section in the first and third cases, we shall use the expression for the scattering cross section per one electron in the case when $\hbar\omega_0 < E_g$:

$$\sigma_{\Delta l=2} \approx \left(\frac{e^2}{mc^2}\right)^2 \left(\frac{\hbar\omega_c}{E_g}\right)^2 \quad cm^2/sr, \tag{38}$$

$$\sigma_{\Delta l=0,\ \Delta s=1} \approx \left(\frac{e^2}{m_s c^2}\right)^2 \left(\frac{\hbar\omega_0}{E_g}\right)^2 \quad cm^2/sr, \tag{39}$$

where m_s is the spin effective mass, defined by

$$m_s = \frac{2m_0}{|g|}. \tag{40}$$

In the case of electrons in n-type InSb the measured scattering cross sections are in agreement with Eqs. (38) and (39) and they amount to [119, 120]:

$$\sigma_{\Delta l=2} \approx \sigma_{\Delta l=1} \approx 10^{-24} \quad cm^2/sr,$$

$$\sigma_{\Delta l=0,\ \Delta s=1} \approx 10^{-23} \quad cm^2/sr.$$

It is worth noting that the cross sections for the scattering by plasmons and one-particle excitations in crystals are considerably larger than the Thomson cross section for the scattering by free electrons. The largest cross section is exhibited by the spin flip process. Consequently, all three processes discussed above can be used to achieve tunable Raman scattering.

B. Estimate of the Threshold. The stimulated emission is obtained if the amplification in a crystal is sufficient to balance out the optical losses in the resonator. The rate of rise of the number of photons in a given optical mode of the Stokes component of the Raman scattering is [121, 122]

$$g_s = \frac{16\pi^2 c^2 (S_0/ld\Omega)}{\hbar\omega_s^3 n_0 n_s (\bar{n} + 1)\Gamma} \mathscr{I} \quad cm^{-1}, \tag{41}$$

where $S_0/l d\Omega$ is the Raman scattering efficiency $(\text{cm}^{-1} \cdot \text{sr}^{-1})$; $\mathscr{J}$ is the intensity of the incident radiation (W/cm^2); ω_s is the Stokes frequency of the scattered light; n_0 and n_s are the refractive indices at the wavelength of the incident and scattered light; $(\bar{n} + 1)$ is the Boltzmann factor, in which

$$\bar{n} = [\exp(g\beta H/k_0 T) - 1]^{-1}; \tag{42}$$

Γ is the total width of the spontaneous scattering line. The Raman scattering efficiency is defined by

$$\eta = (S_0/l d\Omega) = \sigma n_e f(F, H), \tag{43}$$

where n_e is the electron density; $f(F, H)$ is the factor which allows for the electron statistics in a magnetic field. In the case of the Raman scattering by Landau levels we have $f(F, H) \leq 1$, whereas for the scattering by plasmons we have $f(F, H) = 1$. We can see from Eqs. (41) and (43) that in order to achieve a high gain g_s, we must ensure that the cross section σ is large and that the spontaneous scattering line is narrow.

The spin-flip process is characterized not only by the largest scattering cross section but also by the narrowest spontaneous scattering line. Detailed measurements carried out on InSb [120], InAs [117], and PbTe [123] indicated that the line width for the transitions $\Delta l = 1$ and $\Delta l = 2$ was $\Gamma \approx 10\text{-}30 \text{ cm}^{-1}$ whereas for the transitions $\Delta l = 1$, $\Delta s = 1$ it was $\Gamma \lesssim 2 \text{ cm}^{-1}$. Measurements of the width of the plasmon scattering line of GaAs [114-116] and InAs [117] gave $\Gamma \approx 10\text{-}20 \text{ cm}^{-1}$, which was again greater than for the spin-flip process. The line corresponding to the transition $\Delta l = 1$, $\Delta s = 1$ was so narrow because the collisions between electrons made no contribution to the line width, which was primarily due to the conduction-band nonparabolicity.

We shall now estimate the gain in the case of the spin-flip Raman scattering in InSb under the following conditions: $n_e \approx 1.8 \times 10^{16} \text{ cm}^{-3}$, $T = 20°K$, $H = 50 \text{ kOe}$, $|g| \approx 42$, $\lambda_s = 12 \mu$, $\lambda_0 = 10.6 \mu$, $n_0 = n_s = 4$, $\Gamma = 2 \text{ cm}^{-1}$. In this case we have $f(F, H) \approx 1$ and $(\bar{n} + 1) \approx 1$ because $g\beta H \gg k_0 T$, and the gain is

$$g_s \approx 1 \cdot 10^{-5} \cdot \mathscr{J} \text{ cm}^{-1}. \tag{41a}$$

If the pumping rate is $\mathscr{J} \approx 4 \times 10^5 \text{ W}/\text{cm}^2$, the absorption by free carriers is $\alpha \approx 2 \text{ cm}^{-1}$ (in $H = 50 \text{ kOe}$), the length of a crystal is $l = 0.5 \text{ cm}$, and the reflection coefficient is $R = 36\%$, the gain due to a single pass in the resonator is sufficient to exceed the losses due to absorption and one reflection from the end: $R \cdot \exp[(g_s - \alpha) l] = 0.36 \cdot \exp[(4 - 2) \cdot 0.5] \approx 1$.

Thus, focusing of the laser pump radiation of $\sim 1 \text{ kW}$ power into a spot of $\sim 10^{-3} \text{ cm}^2$ area should ensure that the threshold is exceeded by a larger margin.

This estimate applies to the nonresonance case. If $\hbar\omega_0 \to E_g$, the threshold is much lower because the scattering matrix element includes terms of the type

$$\frac{\omega_s}{\omega_0} \sum_k \left| \frac{1}{\mathscr{E}_{c\downarrow}(\mathbf{k}) - \mathscr{E}_v(\mathbf{k}) - \hbar\omega_0} \right|^2, \tag{44}$$

where $\mathscr{E}_{c\downarrow}(\mathbf{k})$ is the energy of a spin sublevel in the conduction band, which depends on the magnetic field; $\mathscr{E}_v(\mathbf{k})$ is the energy in the valence band.

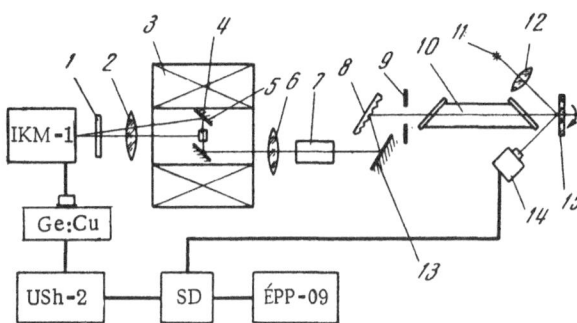

Fig. 41. Schematic diagram of the apparatus used in studies of stimulated Raman scattering: 1) BaF_2 lens; 2) KRS-5 lens; 3) solenoid; 4) sample; 5) rotatable mirrors; 6) cutoff; 7) attenuator; 8) echelette grating; 9) diaphragm; 10) laser; 11) incandescent lamp; 12) lens; 13) mirror; 14) photodiode; 15) rotating mirror.

Fig. 42. Emission spectrum of a CO_2 laser ($\lambda_0 = 10.595\ \mu$) and stimulated Raman scattering by the conduction electrons in indium antimonide subjected to a magnetic field of 50 kOe.

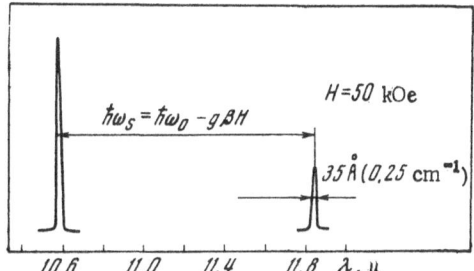

§2. Stimulated Raman Scattering of Light Accompanied by Spin Flip in Indium Antimonide

We investigated and studied the stimulated Raman scattering by the conduction electrons in n-type InSb crystals under transverse geometry conditions (the direction of the scattered light was perpendicular to the pump radiation: $k_s \perp k_0$) [25]. The process responsible for the scattering of light was the reversal (flip) of the electron spin oriented along a static magnetic field by a strong electromagnetic wave (transition $\Delta l = 0$, $\Delta s = 1$).*

The investigated crystals had an electron density of $(1.5-1.8) \times 10^{16}$ cm^{-3} and a mobility of about 10^5 $cm^2 \cdot V^{-1} \cdot sec^{-1}$. A O-switched CO_2 laser, tuned to $\lambda_0 = 10.591\ \mu$, was used as the pump source. The pump radiation power was ~ 1 kW per pulse.

The apparatus employed is shown schematically in Fig. 41. Plane-polarized laser radiation was focused by a BaF_2 lens with a focal length 270 mm. This radiation was directed by a rotatable mirror onto the end of a sample at right angles to the magnetic field so that the electric vector of the pump radiation was parallel to the applied field [109]. The intensity of the pump radiation was $\sim 10^6$ W/cm^2 and it was concentrated in a focal spot of 300 μ diameter. Crystals of $2 \times 2 \times 5$ mm dimensions were located inside the superconducting solenoid described earlier. The long edges of the samples (5 mm) were perpendicular to the magnetic field. The ends of the samples and the resonator faces were polished and they were parallel to within 1'. The temperature of a sample was $\sim 20°K$. The high-power pump radiation was separated by placing an interference cutoff filter in front of the monochromator. The pump radiation could be attenuated by a set of Ge or Si plates or by the use of a cell filled with diethyl ether vapor.

Figure 42 shows the spectra of the laser pump radiation and of the stimulated Raman scattering by the conduction electrons in InSb subjected to magnetic field of 50 kOe. The stimulated Raman scattering line ($\lambda_s = 11.84\ \mu$) was found to be shifted to the Stokes region. Its frequency ω_s, corresponding to the spin flip scattering, varied with the magnetic field in accordance with the law $\omega_s = \omega_0 - g\beta H/\hbar$. The width of this line was $\lesssim 0.25$ cm^{-1} and it was

*The observations of the stimulated Raman scattering in InSb were reported earlier in [23, 24].

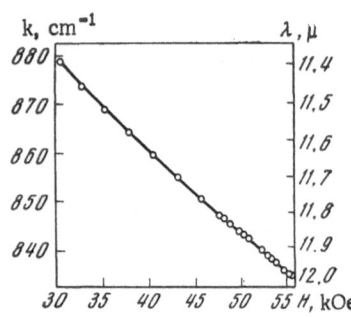

Fig. 43. Dependence of the stimulated Raman emission frequency on the magnetic field.

governed by the transfer function of the spectrometer. The spectrometer resolution allowed us to record two neighboring modes which were generated in a resonator 2 mm long and separated by a frequency interval ~ 0.62 cm^{-1}. However, we always observed only one emission line whose wavelength could be tuned continuously by varying the magnetic field.

The dependence of the frequency of the stimulated Raman emission on the magnetic field is plotted in Fig. 43. When the magnetic field was increased from 30.6 to 56 kOe, the wavelength of the stimulated Raman line changed from 11.4 to 12 μ. The dependence of the frequency on the magnetic field was smooth but it deviated from linearity. The upper limit of the tuning range was limited by the highest magnetic field available. The stimulated Raman emission was not observed below 30 kOe. In fields between 30 and 40 kOe the intensity of the stimulated Raman emission rose severalfold and it fell slightly in the range H > 50 kOe.

The measurements of the pulsed Raman emission power at H = 38 kOe gave a maximum value $P_{max} \approx 0.8$ W, which corresponded to a conversion efficiency of 0.08%.

The stimulated Raman emission was polarized at right-angles to the magnetic field. The degree of polarization was $\sim 95\%$.

Apart from the crystals mentioned above, we investigated also the stimulated Raman emission in crystals with carrier densities of 0.56×10^{16}, 0.74×10^{16}, 3.4×10^{16}, and 3.7×10^{16} cm^{-3}. The stimulated Raman effect was not observed in these crystals throughout the investigated range of magnetic fields (up to 56 kOe) even when the pump radiation intensity had its maximum value of 9×10^5 W/cm^2 (after allowance for reflection).

§ 3. Discussion of Results

The observed radiation was of the stimulated type because the line width (0.25 cm^{-1}) was much less than $k_0 T \approx 2$ meV (~ 16 cm^{-1}). Moreover, the stimulated nature of the emission was indicated by the threshold dependence of the Raman emission intensity on the pump power. When the pump power was below the threshold ($\mathscr{I}_{th} \approx 6 \times 10^5$ W/cm^2 inside the crystal), the intensity of the scattered light measured with a parabolic mirror varied approximately linearly with the pump power $\mathscr{I}$. In the range $\mathscr{I} \gtrsim \mathscr{I}_{th}$, the intensity of the scattered light rose rapidly (by a factor exceeding 100 for $\mathscr{I}_{max} \approx 9 \times 10^5$ W/cm^2).

A simultaneous analysis of the shape of the Raman and pump light pulses on the screen of an oscillograph demonstrated that the strong Raman emission lagged somewhat behind the pump pulse so that we found that $\mathscr{I}_{max} = 0.7 \mathscr{I}_{max}$. This was also evidence of the threshold nature of the observed phenomenon. When the duration of the pump pulses was ~ 0.3 μsec, the duration of the Raman emission pulses was about 100 nsec.

However, it was not clear why the stimulated Raman emission frequency varied smoothly (within the limits of the spectral resolution of the instruments employed) with the magnetic field and why the spectra did not exhibit the mode structure of the resonator. This could be due to superscattering (an analog of superluminescence, when one transit through the resonator was sufficient to reduce considerably the width of the emission line) or it was due to

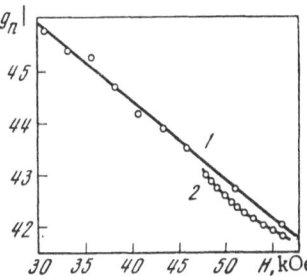

Fig. 44. Dependence of the g factor of electrons in InSb on the magnetic field, plotted for two electron densities $n \times 10^{-16}$ (cm^{-3}): 1) 1.5; 2) 1.8. $T \approx 20°K$.

the complex structure of the electromagnetic field in the resonator. This question deserves further study.

Since the dependence of the scattered-light frequency on the magnetic field obeys the law $\omega_s = \omega_0 - g\beta H/\hbar$, it follows that the excitation responsible for the scattering of light is the transition of an electron from a lower spin Landau sublevel in the conduction band to an upper level ($\Delta l = 0$, $\Delta s = 1$). In contrast to the recombination radiation, the transfer of an electron is associated with one band and this makes it possible to determine the g factor directly. Since the emission line is very narrow, the precision in the determination of the g factor is considerably higher than that of other known methods.

Figure 44 shows the dependence of the g factor of electrons on the magnetic field applied to InSb with two different electron densities. We can see that a slight increase in the electron density, resulting in a fall of the g factor [97], is manifested clearly in the stimulated Raman spectrum. The reduction in the g factor with increasing magnetic field is due to the band nonparabolicity [34].

We shall now consider the magnetic field threshold below which the stimulated Raman emission is not observed and the reason for the absence of such emission in the case of samples with nonoptimal electron densities. We shall do this by analyzing Eq. (41) and assuming that the width of the spontaneous scattering line Γ and the refractive indices n_0 and n_s are independent of the field. The shift of the stimulated Raman frequency in the direction of long wavelengths, which is observed when the magnetic field is increased, tends to increase the gain g_s but this increase is slight ($\sim 17\%$) in the investigated tuning range between 11.4 and 12.0 μ. Moreover, we may assume that $n + 1 \approx 1$, since $g\beta H > k_0 T$ is given in fields $H \approx 10$ kOe. Therefore, the gain in the resonator is governed primarily by the scattering efficiency η and the pump intensity $\mathcal{J}$. Since η is proportional to the electron density n_e, this density must be increased in order to obtain a higher value of the gain. However, it is sensible to raise n_e in a given magnetic field only in the quantum limit case, i.e., when the Fermi level of electrons is located between the spin sublevels of the lowest Landau level. This is reflected in Eq. (43), which gives the scattering efficiency in terms of the factor $f(F, H)$, whose value is unity in the quantum limit. However, if the density is fixed, a reduction of the field may result in the violation of the quantum limit condition. Then, $f(F, H) < 1$, and the scattering efficiency falls.

This phenomenon is illustrated in Fig. 45, which gives the energy positions of the spin-split Landau levels in the conduction band of indium antimonide as a function of the electron wave vector $\mathbf{k}_z(\mathbf{k}_z \| \mathbf{H})$ for two values of H. In each case the position of the Fermi level is given for three electron densities of 5.6×10^{15}, 1.5×10^{16}, and 3.7×10^{16} cm^{-3}, each of which also depends on the magnetic field. The calculated results plotted in this figure are based on Eqs. (12), (10), and (10a). We can see that when the electron density is 5.6×10^{15} cm^{-3} the quantum limit is satisfied even by fields of 20 kOe, but when the density is 3.7×10^{16} cm^{-3}, this limit is not reached even in fields of 50 kOe. Hence, we may conclude that the stimulated Raman emission is not observed for samples with low and high carrier densities when the

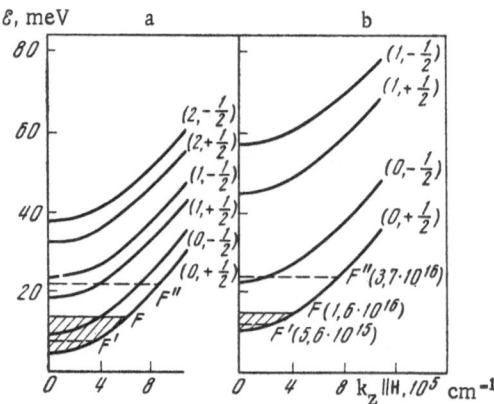

Fig. 45. Energy positions of two spin-split Landau levels in the conduction band of indium antimonide plotted as a function of the electron wave vector in the direction of a magnetic field H = 20 kOe (a) and H = 50 kOe (b). The position of the Fermi level for an electron density of 1.6×10^{16} cm^{-3} is represented by a continuous line F; the positions of the Fermi level for 5.6×10^{15} and 3.7×10^{16} cm^{-3} are represented by dashed lines F' and F"; energies are measured from the position of the bottom of the conduction band in H = 0.

pump intensity is 9×10^5 W/cm^2 because of the fall of the scattering efficiency. In the case of samples with low electron densities the number of carriers is simply too small, whereas in the case of samples with high electron densities the quantum limit is not reached. It should be possible to obtain the stimulated Raman emission in the case of samples with high electron densities if the pump intensity is made sufficiently high and the magnetic field is sufficiently strong.

It is clear from Fig. 45a that in the range of intermediate (optical) electron densities in a field H = 20 kOe the upper spin sublevel (l = 0, s = −1/2) is only partly filled and since the scattering results in the transfer of electrons from the lower sublevel (l_0 = 0, s = +1/2) to the upper one, only those electrons which are close to the Fermi level can participate in the scattering process. If the magnetic field is increased to 50 kOe and the carrier density remains constant, the upper sublevel becomes completely empty and we have $f(F, H) = 1$. The stimulated Raman emission from samples with intermediate carrier densities in fields ~30 kOe indicates that the quantum limit is reached (this limit can be regarded as the magnetic threshold for the stimulated Raman emission). The thermal broadening of the Fermi level in stronger fields causes the intensity of the stimulated Raman emission to rise somewhat. The subsequent slow fall of the intensity beyond the quantum limit results from a slight reduction in the scattering cross section in magnetic fields, which is predicted theoretically [109] for the $\hbar\omega_0 \ll E_g$ case [see Eq. (39)].

The gain in the resonator is affected not only by the factors considered above but also by the resonator losses. They are composed of the reflection losses which are due to the resonator walls (assumed not to be coated with antireflection films) and which are independent of the magnetic field, and of the absorption losses due to free carriers, which are dependent on the magnetic field because $\alpha \propto [\lambda_s (H)]^2$, where λ_s rises with the magnetic field. Naturally, the use of samples with higher carrier densities or reduction in the carrier density in a sample due to the two-photon absorption would raise the resonator losses and the threshold.

The relatively low values of the conversion efficiency obtained experimentally are due to the fact that the pump power is insufficient, at least to ensure saturation [110].

Thus, a study of the Raman scattering of light by Landau levels in semiconductors yields important information on the energy structure. In particular, an investigation of the spin flip process allows us to determine very accurately the dependence of the g factor on the magnetic field, carrier density, and temperature. The problem is simplified by the fact that electron transitions occur within one band.

On the other hand, the stimulated spin-flip Raman scattering in InSb is of great practical importance for infrared applications because this scattering is characterized by a relatively high power output, very narrow emission line, and continuous tuning in a wide range. It seems that the conversion efficiency may be increased considerably by the use of IV-VI semiconductor compounds or of ternary compounds such as $Pb_{1-x}Sn_xSe$, $Pb_{1-x}Sn_xTe$, and $Cd_{1-x}Hg_xTe$. These materials have high values of the g factor and are also interesting because of the possibility of variation of the forbidden-band width. This circumstance should make it possible to adjust the values of E_g of the semiconducting material to the CO_2 laser emission frequency so as to achieve a resonance enhancement of the efficiency of the stimulated Raman effect.

The important advantages of the InSb spin-flip Raman laser and its promising applications have stimulated the appearance of quite a few papers although not more than two years have passed since the construction of the first such lasers. A resonance enhancement of the scattering in the case of pumping with CO laser radiation was reported in [24, 124-127], an increase in the stimulated Raman emission intensity in the case of antireflection-coated samples was described in [128], and the stimulated emission of the second Stokes component [127, 129] and the first anti-Stokes component [130] were observed. Record output powers were obtained when a CO_2 laser, operating at atmospheric pressure, was used as the pump source [131]. The magnetic threshold was lowered to 400 Oe by reducing the carrier density under resonance pumping conditions [132]. The heterodyne method was used in the measurement of the width of the emission line of a cw Raman laser [133]. This width was found to be ~1 kHz. The tuning range was extended quite considerably [110, 134, 135]. This laser was used to record the first high-resolution absorption spectra of NH_3 [134], H_2O [136], and NO [137] vapors.

CHAPTER V

INFLUENCE OF PRESSURE ON RADIATION EMITTED FROM LEAD SELENIDE AND GALLIUM ARSENIDE SEMICONDUCTOR LASERS

§1. Emission Spectra of PbSe Lasers

The investigations were carried out on diodes [138] similar to those which were employed in magnetooptic measurements (see § 3 in Chap. III). Since the intensity of the spontaneous radiation emerging from a high-pressure chamber was very low when the pulse duration was 1-3 μsec (this intensity exceeded only slightly the noise level of the recording system) the main measurements of the spectra were carried out under stimulated emission conditions at liquid-nitrogen temperature.

Typical emission spectra of a PbSe laser obtained at various pressures are shown in Fig. 46. The mode structure and the pressure dependence of the emission wavelength were clearly visible. Since the injection current was kept the same in recording of all these spectra,

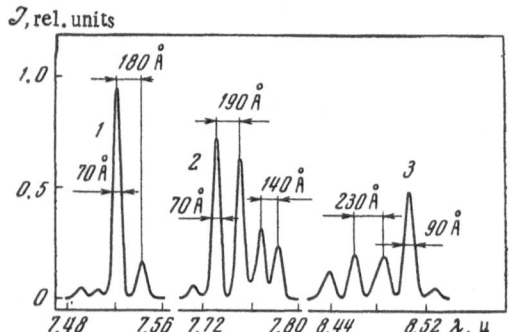

Fig. 46. Emission spectra of a PbSe laser at 77°K and different pressures P (kbar): 1) 0.51; 2) 1.8; 3) 2.75. The injection current was 1.5 times higher than the threshold value.

their changes were entirely due to the pressure. For example, at a pressure P = 0.51 kbar the laser action was concentrated mainly in one central mode, whereas at P = 1.08 kbar the spectrum became asymmetric and it spread in the direction of shorter wavelengths but at P = 2.75 kbar it spread in the direction of longer wavelengths. This was due to the fact that at some pressures in the resonator several different modes with different Q factors were excited (see, for example, [139]) and this was confirmed by the presence of nonequidistant modes in the emission spectrum obtained at P = 1.08 kbar.

The separation between the equidistant modes $\Delta\lambda$ varied approximately proportionally to the square of the emission wavelength, in accordance with Eq. (22). When the resonator length was l = 0.43 mm and the pressure was increased from atmospheric to 5 kbar, the separation $\Delta\lambda$ increased from 160 to 300 Å. The effective refractive index was

$$\left(n - \lambda \frac{dn}{d\lambda}\right) \approx 6.$$

The width of the laser modes generated using short current pulses (0.8-1 μsec) was usually governed by the transfer function of the spectrometer (Fig. 46). When the maximum resolution of the monochromator was used, the measured width was less than 40 Å ($\sim$ 0.7 cm^{-1} at λ = 7.7 μ). When the pulse duration was increased to 3 μsec, the temperature of the crystal rose significantly, which resulted in a drift of the emission modes during a pulse and in their broadening to 120 Å.

The laser emission wavelength increased with pressure (Fig. 46). In the 0-5 kbar range it rose from 7.34 to 9.63 μ (Fig. 47). The nature of this pressure tuning depended on the rate of carrier injection. In the multimode emission case an increase in the pressure resulted in the excitation of the long-wavelength modes and quenching of the short-wavelength ones. Determination of the pressure dependence of the integrated intensity of the multimode radiation revealed intensity modulation (about 15%). When the emission was close to the single-

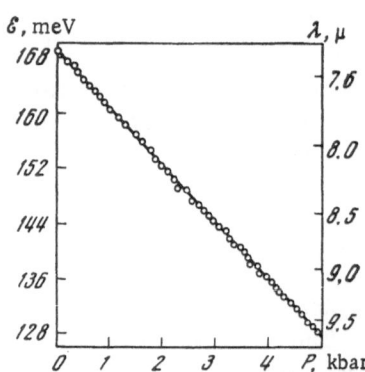

Fig. 47. Pressure dependence of the energy of photons emitted by a PbSe laser at 77°K.

mode case (i.e., when the current was only slightly higher than the threshold value), an increase of the pressure resulted in the transfer of energy from the short- to the long-wavelength mode. However, in a certain range of pressures the stimulated emission disappeared completely. In this range the laser action was restored by increasing the current.

The photon energy in Fig. 47 was measured at the center of gravity of the modes in the laser emission spectrum at a given pressure. The wavelength corresponding to the center of gravity was found from the expression

$$\lambda_{cg} = \frac{\sum_{i}^{N} \mathcal{J}_i \lambda_i}{\sum_{i}^{N} \mathcal{J}_i}, \tag{45}$$

where λ_i and $\mathcal{J}_i$ are, respectively, the wavelength and intensity of the i-th mode; N is the number of modes in the spectrum.

It is clear from Fig. 47 that the experimental points fit well a straight line up to 5 kbar. Hence, it follows that the pressure dependence of the effective mass is weak, at least in the range of pressures under investigation. The slope of the straight line in Fig. 47 corresponds to a negative pressure coefficient of the forbidden band width E_g whose value is $(\partial E_g / \partial P)_{77^\circ K} = (-8.2 \pm 0.3) \cdot 10^{-6}$ eV/bar.

This pressure coefficient is in good agreement with the value $(\partial E_g / \partial P)_{77^\circ K}$ deduced from electrical measurements in [140] and from the photovoltaic effect [141] in PbSe.

As mentioned in Chap. I, apart from the direct influence on the band structure, the application of pressure alters also the refractive index and the dimensions of a crystal. These effects give rise to a measurable shift of the laser modes even over a narrow range of pressures. This range is governed by the values of the pressure at which a given mode is excited and quenched. The emission spectra of a PbSe laser obtained at slightly differing pressures are shown in Fig. 48. We can see clearly that the emission spectrum shifted in the direction of longer wavelenths with rising pressure.

The shifts of the frequencies of the individual PbSe laser modes with pressure rising in small steps between 1.3 and 2.2 kbar is shown in Fig. 49. The dashed line in this figure is the pressure dependence of E_g. The results plotted in Fig. 49 were obtained using a current which was twice as high as the threshold value, so that between four and seven modes appeared simultaneously in the emission spectrum. It is clear from Fig. 49 that when the pressure was increased the energy of each mode decreased. The range of pressures in which a particular mode was observed amounted to about 300 bar. In this range the reduction in the mode energy was about 0.6 meV, which corresponded to a continuous tuning of the emission

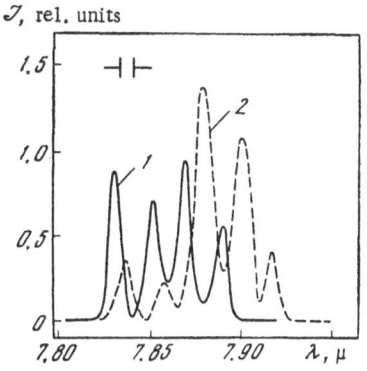

Fig. 48. Emission spectra of a PbSe laser obtained for slightly differing pressures P (kbar): 1) 1.32; 2) 1.37.

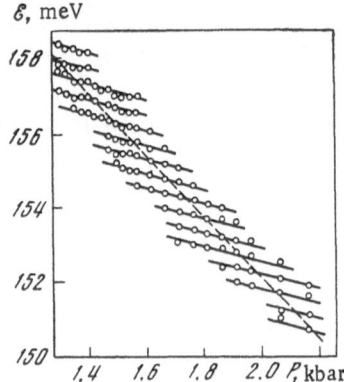

Fig. 49. Pressure-induced shifts of the emission frequencies of individual modes of a PbSe laser.

frequency in the range of ~ 150 GHz (or ~ 5 cm^{-1}). The pressure coefficient of the mode energy was $(\partial \mathscr{E}_{mode}/\partial P)_{77^\circ K} = (-2.1 \pm 0.6) \cdot 10^{-6}$ eV/bar, which was approximately one-quarter of the pressure coefficient of the forbidden width of PbSe. The pressure coefficient of the mode energy was in good agreement with the results reported in [19].

The experimentally determined pressure coefficients of the emission wavelength and mode separation could be combined with Eq.(20a) in a calculation of the pressure coefficient of the refractive index of the active medium:

$$\left(\frac{\partial n}{\partial P}\right)_\lambda = \frac{\lambda}{2l\Delta\lambda} \cdot \frac{d\lambda}{dP} \approx 5 \cdot 10^{-5} \text{ bar}^{-1}.$$

We must point out an important feature of the observed continuous tuning of the emission frequency. The range of this tuning was ~ 150 GHz (~ 280 Å), which was considerably greater than the mode separation at the pressures employed, i.e., when the pressure was varied by 300 bar, each mode shifted by an amount exceeding the mode separation. Thus, for the selected injection rate and length of a crystal, it was possible to achieve a continuous frequency tuning throughout the investigated pressure range.

The interval of the continuous pressure tuning of each mode (150 GHz) was three times as large as the frequency tuning achieved for the ternary compound Pb$_{1-x}$Sn$_x$Te by varying the temperature [26].

We used a IPP-12 polarizer in measurements of the polarization of the radiation emitted by the investigated PbSe laser. Two typical spectra, obtained at 2.5 kbar, are shown in Fig. 50. In this figure the angle between the plane of polarization of the polarizer and the p—n junction plane is denoted by θ. We can see that the polarization of different modes was different. Preferential polarization was possible in the p—n junction plane and in a plane perpendicular to the p—n junction. The degree of polarization of each individual mode usually remained constant in a fairly narrow range of pressures.

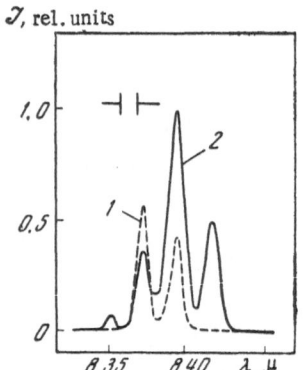

Fig. 50. Polarization of the emission modes of a PbSe laser operated at a pressure P = 2.5 kbar: 1) $\theta = 0^\circ$; 2) $\theta = 90^\circ$.

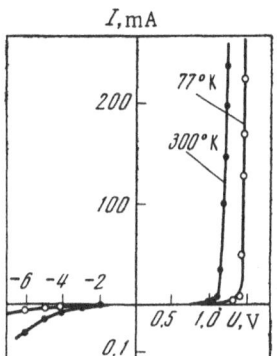

Fig. 51. Current–voltage characteristics of a GaAs diode at 77 and 300°K.

§2. Emission Spectra of GaAs Lasers

Laser diodes were prepared from n-type tellurium-doped gallium arsenide single crystals. The carrier density was usually $(1-5) \times 10^{18}$ cm^{-3} and a p–n junction was produced by the diffusion of zinc. The reflecting faces of the resonator were the natural (100) or (110) crystallographic planes, revealed by cleaving.

The current–voltage characteristics of a gallium arsenide diode, obtained at 77 and 300°K are shown in Fig. 51. The value of r in the argument of the exponential function describing the forward current at 77°K was 2.

The emission spectra under pressure were determined at 77°K. The recombination radiation emitted at high injection levels was dominated by transitions between the conduction band and a zinc impurity level [142]. The laser action occurred in these transitions at atmospheric pressure and the wavelength of the resultant radiation was $\lambda \approx 0.84\ \mu$.

The pressure dependences of the positions of the gain profiles of GaAs lasers were obtained using two pressure-transmitting media: one of them was helium and the other was a frozen mixture of oil and kerosene. In the former case the maximum working pressure was 2.65 kbar, whereas in the latter case it was about 9 kbar.

The laser emission spectra obtained at two pressures are shown in Fig. 52. It is evident from this figure that the emisssion wavelength became shorter with increasing pressure. The temperature shift, due to a slight change in the current density, was small (it will be considered later). The use of a quasihydrostatic pressure-transmitting medium did not damage the sample and the second spectrum in Fig. 52 consisted of a narrower emission line. The line width was set by the transfer function of the IKM-1 monochromator. The monochromator resolution (~ 2.5 Å) was insufficient to observe the mode structure.

The dependence of the energy of the photons emitted by a GaAs laser on the hydrostatic pressure applied at 77°K is shown in Fig. 53. In this case, helium was used as the pressure-

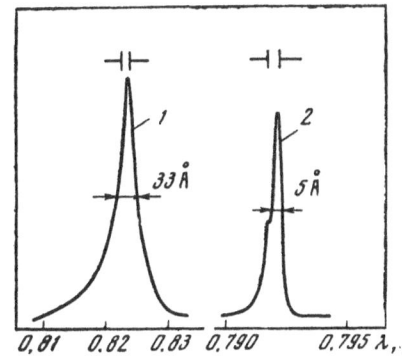

Fig. 52. Emission spectra of a GaAs laser: 1) 2.5 kbar (j = 4 × 10^3 A/cm^2); 2) 7.95 kbar (j = 6 × 10^3 A/cm^2).

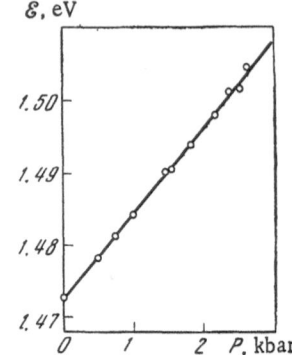

Fig. 53. Dependence of the energy of the GaAs laser photons on the hydrostatic pressure applied at 77°K.

transmitting medium. The experimental points fitted well a straight line corresponding to the
following pressure coefficient of the forbidden-band width of gallium arsenide: $(\partial E_g / \partial P)_{77°K} =$
$(11.5 \pm 0.6) \cdot 10^{-6}$ eV/bar.

The value of this pressure coefficient was in good agreement with the results given in
[18].

We found that the true pressure in a frozen oil and kerosene mixture could not be determined by standard methods. Therefore, we found local pressures exceeding 3 kbar by
measuring the wavelength shift and applying the pressure coefficient found above (Fig. 54).
This extrapolation was justified in [18] up to pressures of 8 kbar. It is clear from Fig. 54
that when the pressure was increased from atmospheric to 8.8 kbar, the emission wavelength
decreased from 0.84 to 0.79 μ.

The laser emission wavelength obtained at a fixed pressure was also a function of the
injection current: it shifted in the direction of longer wavelengths when this current was increased. This was due to the heating of the crystal [142], which resulted in a considerable
reduction in the forbidden-band width at high currents. The consequent shift of the emission
wavelength amounted to 3 Å/A for current densities from 10^3 to 10^4 A/cm^2 (the pulses employed
were of 2 μsec duration and 200 Hz repetition frequency). The Burstein shift in the direction
of shorter wavelengths was masked by the greater shift due to heating and it was observed only
at sufficiently low current densities, when the heating effect was small.

We assumed that the shift was solely due to the heating and thus found the minimum
temperature rise, which was $\Delta T = 10$ deg K. Allowance for the filling of the allowed bands
with carriers simply increased this value of ΔT.

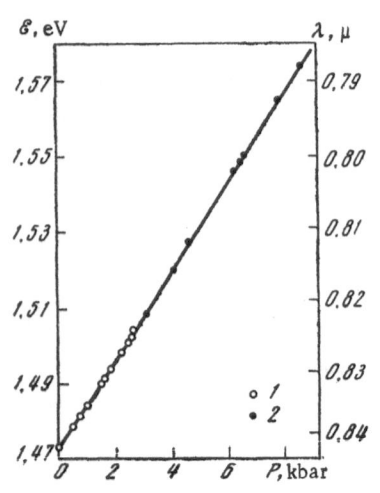

Fig. 54. Dependence of the energy of the GaAs
laser photons on the hydrostatic pressure applied
at 77°K. Pressure-transmitting medium: 1) helium;
2) oil + kerosene.

§ 3. Discussion of Results

As pointed out in § 2, Chap. I, measurements of the pressure coefficients of the forbidden band width E_g could be used to estimate the contribution of the lattice expansion to the temperature dependence of E_g. The coefficients obtained for GaAs and PbSe were basically similar to those listed in Table 5. Therefore, the contribution in question was of the same order of magnitude so that at 77°K the electron–phonon interaction was responsible for about half of the temperature coefficient of E_g of PbSe and four-fifths in the case of GaAs.

We shall now consider the tuning characteristics of the investigated lasers. It follows from our measurements that the widest tuning range is obtained by applying pressure. For the sake of comparison, we shall compare the relative ranges of tuning of the emission wavelengths of the investigated lasers $\Delta\lambda/\lambda_{av}$ achieved using magnetic fields and pressures. In magnetic fields the tuning range is 1.9% for InAs, 5.4% for InSb, 5.1% for InSb Raman laser, and 2.4% for PbSe. The pressure tuning range is 6.1% for GaAs and 27% for PbSe. Thus, the widest tuning range is observed for lead selenide. It should be noted that the highest hydrostatic pressure of 5 kbar used in our measurements is limited by the available windows and capillaries. If diamond windows and different capillaries, capable of withstanding higher pressures, are employed, it should be possible to reach 14 kbar or higher pressures using helium as the transmitting medium at 77°K [19]. In this case the tuning range would be considerably wider.

It is worth comparing this method of frequency tuning with the pressure tuning of lasers made of other compounds. Table 7 gives the wavelength tuning ranges of some semiconductor lasers expected for pressures up to 20 kbar at 77°K. We can see that the tuning ranges taken together extend from about 1 to 100 μ.

The maximum emission wavelength of PbSe is rather theoretical. This is due to the fact that at long wavelengths we may reach the plasma frequency at least in one of the regions of a p–n junction. In this case the laser action may disappear or some special features may be observed. The plasmon wavelength in lead selenide along the [100] direction is about 30 μ for a carrier density of 2×10^{18} cm^{-3}. Even if this limit is shifted toward longer wavelengths by reducing the carrier density, the recombination radiation may be strongly absorbed by phonons (in the region of 40 μ due to the two-phonon processes and in the region of 80 μ in the case of one-phonon processes).

The injection lasers investigated by us were operated at liquid-nitrogen temperature in the pulsed regime. This means that the width of the emission line was fairly wide because of the temperature drift of the mode during each pulse. For example, in the case of lead selenide we carried out special measurements with a scanning Fabry–Perot infrared interferometer [143] and we found that the line width was less or equal to 10 A when the pulse duration

TABLE 7. Expected Pressure-Tuning Range of
Some Semiconductor Lasers (up to 20 kbar at
77°K)

Material	Wavelength tuning range, μ	Remarks
GaAs	0.73—0.84	
InP	0.82—0.90	
GaSb	1.4—1.6	Up to 8 kbar
InAs	2.3—3.2	
InSb	2.4—5.5	Spontaneous radiation
PbS	4—10	
PbTe	6—20	
PbSe	7.3—100	

TABLE 8. Comparison of Spectral Brightness of Investigated
Lasers and Black Radiators at Similar Wavelengths

Parameter	GaAs	InAs	InSb	PbSe	InSb Raman laser
λ, μ	0.84	3.2	5.3	8.5	12
P, W	0.1	0.05	0.05	0.006	1
S, cm^2	$3 \cdot 10^{-5}$	$2 \cdot 10^{-5}$	$5 \cdot 10^{-5}$	10^{-4}	10^{-2}
$\delta\lambda$, μ	$5 \cdot 10^{-5}$	10^{-3}	$1.2 \cdot 10^{-3}$	10^{-3}	$3.5 \cdot 10^{-3}$
$d\Omega$, sr	$7 \cdot 10^{-4}$	$4 \cdot 10^{-3}$	$8 \cdot 10^{-3}$	10^{-2}	$1.5 \cdot 10^{-3}$
B_λ, $W \cdot cm^{-2} \cdot \mu^{-1} \cdot sr^{-1}$	10^{10}	$6 \cdot 10^8$	10^8	$6 \cdot 10^6$	$2 \cdot 10^7$
B_λ^{theor}, $W \cdot cm^{-2} \cdot \mu^{-1} \cdot sr^{-1}$	$1.8 \cdot 10^4$	0.68	0.19	$4 \cdot 10^{-2}$	$1.2 \cdot 10^{-2}$

was 1 μsec. No significant reduction in the line width could be expected from a reduction in
the pulse duration.

Nevertheless, it would be interesting to estimate the spectral brightness of tunable lasers
and to compare them with the classical infrared radiation sources.

The brightness B_λ for a single mode will be defined as follows:

$$B_\lambda = \frac{W}{S\delta\lambda d\Omega}, \tag{46}$$

where W is the power of a single mode, S is the area of the radiating part of the p−n junction
surface, $\delta\lambda$ is the width of the emission line, and $d\Omega$ is the solid angle.

Table 8 gives the calculated values of the spectral brightness for single-mode emission
from the investigated lasers. The line width was assumed to be equal to the transfer function
of the spectrometer. The brightness estimates were obtained using the experimental values
of the output power and angular divergence. The data for GaAs and InSb were taken from the
literature.

This table gives also the value of the spectral density B_λ^{theor} calculated for λ = 0.84
μ emitted by a xenon lamp source ($T_e \approx 3 \times 10^4$ °K) and for λ = 3.2, 5.3, 8.5, and 12 μ emitted
by a Globar source (T = 1700°K). These theoretical values of the brightness were calculated
on the assumption that both sources are black radiators.

We can see that the spectral brightness of the GaAs laser is approximately six orders
of magnitude higher than that of the xenon lamp and the difference is eight to nine orders of
magnitude in the other cases. However, the emitting area of a semiconductor laser is usually
two or three orders of magnitude smaller than the area of a slit filled with light from a globar.
Therefore, the actual advantage in respect of the light flux is correspondingly smaller. If in
this comparison with the classical sources we also allow for the pulsed operation of the lasers,
we find that the advantage should be reduced by another two orders of magnitude. Nevertheless,
the net advantage is still high. In the case of cw lasers the advantage is particularly large

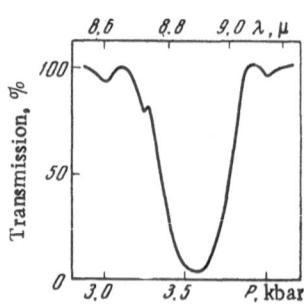

Fig. 55. Absorption spectrum of single-crystal silicon
at 77°K, recorded using a pressure-tuned PbSe laser.

mainly because the emission line is considerably narrower (by five or six orders of magnitude) and the advantages of pressure- or magnetic-field-tuned lasers are very great.

The operation of a spectrometer coupled to a PbSe laser is demonstrated in Fig. 55 by the absorption spectrum of single-crystal silicon recorded at 77°K by varying the pressure. The width of the absorption band is $\Delta k = 28$ cm^{-1} at $\lambda \approx 8.9$ μ. The width of the multimode laser emission band is less than 10 cm^{-1}. The sample is a silicon window in a high-pressure chamber so that the integrated radiation emerging from this chamber is simply focused on the sensitive area of a detector. The characteristic silicon absorption band at $\lambda = 8.9$ μ [144] is due to vibrations of the Si—O bonds. The intensity of this band is usually employed in estimating the oxygen concentration in a crystal. In the case under consideration the calculated absorption coefficient for a window ~ 0.5 cm thick is ~ 7 cm^{-1}. The width of the absorption band is ~ 28 cm^{-1}, which is about three times as large as the width of the multimode emission band of the PbSe laser. The observed structure is evidently due to fluctuations of the integrated (over all the modes) laser radiation intensity. These fluctuations may reach 15%. It should be noted that the margin of the gain in recording the spectrum exceeded two orders of magnitude.

CONCLUSION

An investigation of the influence of strong magnetic fields on the emission wavelengths of lasers made of narrow-gap semiconductors InAs, InSb, and PbSe and of the influence of high hydrostatic pressures on the emission from GaAs and PbSe lasers has demonstrated that quasicontinuous tuning of the emission frequency is possible in a relatively wide spectral range (Table 9). The quasicontinuous nature of the tuning is due to the multimode structure of the emission spectra, so that the tuning involves the quenching of some modes and excitation of other modes within a gain profile that shifts with increasing magnetic field or pressure. In a narrow frequency range it is possible to tune continuously the emission frequency of individual modes because of the dependence of the refractive index of the active medium on the applied magnetic field or pressure.

The width of modes emitted under pulsed operation conditions from injection lasers is governed by the frequency drift as a result of transient heating of the laser crystal. Therefore, it would be particularly useful to develop cw lasers in order to ensure the emission lines are narrow. Such lasers can be tuned by pressure or magnetic field and they would be of great practical interest in very-high-resolution spectroscopy, heterodyning of light, etc. The main difficulty is the development of the lasers themselves. The necessary magnetic fields and pressures are easily obtained because no particular difficulties would be encountered in the laboratory generation of fields up to 100 kOe and hydrostatic pressures up to 15 kbar at 77°K.

Raman lasers, particularly those operating continuously under resonance scattering conditions, are very promising. The use of ternary lead chalcogenides should make it possible

TABLE 9. Wavelength Tuning Range of Semiconductor Lasers Subjected to Magnetic Fields and Pressures

Material	Magnetic field (0-56 kOe)		Raman laser (30-56 kOe)		Pressure (5 kbar for PbSe and 8.8 kbar for GaAs)	
	λ, μ	$\Delta\lambda/\lambda_{av}$, %	λ, μ	$\Delta\lambda/\lambda_{av}$, %	λ, μ	$\Delta\lambda/\lambda_{av}$, %
GaAs	—	—	—	—	0.79—0.84	6.1
InAs	3.11—3.17	1.9	—	—	—	—
InSb	5.02—5.30	5.4	11.4—12.0	5.1	—	—
PbSe	8.30—8.50	2.4	—	—	7.34—9.63	27

TABLE 10. Some Parameters of Investigated Materials Deduced from Magnetooptic Measurements and from Experiments under Pressure

Material	$T \approx 10°K$			$(\partial E_g/\partial P)_{77°K}$ eV/bar	$(\partial n/\partial H)$, Oe^{-1}	$(\partial n/\partial P)$, bar^{-1}
	m_n/m_0	$|g_n|$	E_g, eV			
GaAs	—	—	—	$+1.15 \cdot 10^{-5}$	—	—
InAs	0.043	8	0.3881	—	$-2.9 \cdot 10^{-7}$ for $H_\perp = 5$ kOe	—
InSb	0.015	42-46, depending on n and H	0.2335	—	$-1.7 \cdot 10^{-6}$ for $H=9$ kOe	—
PbSe	0.042 for $m_n = m_p$ [100]	30 for $|g_n| = |g_p|$ [100]	0.145	$-8.2 \cdot 10^{-6}$	$-1.2 \cdot 10^{-6}$ for $H=30$ kOe	$+5 \cdot 10^{-5}$ for $P=1.8$ kbar

to build high-power monochromatic light sources emitting in the atmospheric window region (8–14 μ).

The reported investigation of the influence of magnetic fields and pressures on the emission from semiconductor lasers has yielded important information on the parameters of the investigated materials (Table 10), including the effective mass of carriers, g factor, forbidden-band width, pressure coefficient of the forbidden-band width, and pressure and field dependences of the refractive index. Measurements of the Raman scattering spectra in magnetic fields and a study of the simulated Raman scattering by Landau levels has yielded information on the behavior of the main parameters near the band edge, which is not always easy to obtain from magnetoabsorption experiments. The spectra of the Raman scattering by Landau levels in semiconductors make it possible to determine very accurately the dependence of the g factor of electrons or holes on the magnetic field, carrier density, and temperature. Such dependences can be obtained at higher carrier densities than in the electron spin resonance case.

The results of the reported investigation can be summarized as follows.

1. The following apparatus was designed and constructed: an optical helium cryostat with a superconducting solenoid capable of generating fields up to 56 kOe; apparatus for optical investigations under hydrostatic pressures at 77°K; a tunable Q-switched CO_2 laser; infrared detectors made of zinc- or copper-doped germanium. This apparatus made it possible to study the infrared radiation emitted by lasers in strong magnetic fields and under high hydrostatic pressures.

2. An investigation of injection lasers made of narrow-gap semiconductors InAs, InSb, and PbSe in strong (up to 56 kOe) magnetic fields demonstrated that the coherent emission wavelength could be tuned quasicontinuously in a relatively wide range (up to 5.4%) by the shift of Landau levels in the applied magnetic field. A continuous shift of the emission frequency was possible in the case of individual laser modes and this shift was due to the dependence of the refractive index of the active crystal on the magnetic field.

3. It was established that the relative shift of the emission frequency of InSb lasers was over twice as large as the shift of the emission of PbSe lasers, in spite of the comparable values of the cyclotron effective masses and g factors. This demonstrated that, apart from these quantities, the frequency shift depended on the features of the energy band structure such as the presence of light and heavy holes in the valence band of InSb.

4. Laser action due to interband transitions of electrons accompanied by spin flip was observed for the first time in the case of PbSe subjected to a strong magnetic field. The split-

ting of the emission line indicated that the g factors of electrons and holes in PbSe were quite different.

5. The emission spectra of InAs, InSb, and PbSe lasers in magnetic fields yielded the effective masses and g factors of carriers near the band edges, and the widths of the forbidden band (Table 1).

6. The stimulated scattering of light by the conduction electrons in indium antimonide was observed for the first time in the Soviet Union. The inelastic scattering process involved the excitation of electrons from the lower spin Landau sublevel to the upper one, i.e., it resulted in the reversal (flip) of the electron spin in the external magnetic field under the action of a strong electromagnetic wave. In this case the range of the continuous frequency tuning in magnetic fields of 30-56 kOe was 11.4-12.0 μ for a line $\leqslant$0.35 Å ($\leqslant$0.25 cm^{-1}) wide. Measurements of the frequency shift of the Raman laser yielded the dependence of the absolute value of the g factor of electrons in InSb on the magnetic field and carrier density in samples with electron densities of $\sim 10^{16}$ cm^{-3}.

7. An investigation was made of the emission from PbSe and GaAs lasers under hydrostatic pressures. It was found that in this case the frequency tuning range was particularly wide (tens of percent).

When a large number of modes was excited in a lead selenide laser, it was possible to achieve a continuous pressure shift of the emission frequency of individual modes over ranges wider than the mode separation. In this way the whole tuning range could be filled quasicontinuously with laser emission frequencies.

The author is grateful to his scientific director A. P. Shotov for discussing the results obtained and to B. N. Matsonashvili for his constant help.

LITERATURE CITED

1. N. G. Basov, B. M. Vul, and Yu. M. Popov, Zh. Eksp. Teor. Fiz., 37:587 (1959).
2. N. G. Basov, O. N. Krokhin, and Yu. M. Popov, Zh. Eksp. Teor. Fiz., 40:1879 (1961).
3. M. G. A. Bernard and G. Duraffourg, Phys. Status Solidi, 1:699 (1961).
4. D. N. Nasledov, A. A. Rogachev, S. M. Ryvkin, and B. V. Tsarenkov, Fiz. Tverd. Tela, 4:1062 (1962).
5. M. I. Nathan, W. P. Dumke, G. Burns, F. H. Dill, Jr. and G. Lasher, Appl. Phys. Lett., 1:62 (1962).
6. T. M. Quist, R. H. Rediker, R. J. Keyes, W. E. Krag, B. Lax, A. L. McWhorter, and H. J. Zeigler, Appl. Phys. Lett., 1:91 (1962).
7. V. S. Bagaev, N. G. Basov, B. M. Vul, B. D. Kopylovskii, O. N. Krokhin, E. P. Markin, Yu. M. Popov, A. N. Khvoshchev, and A. P. Shotov, Dokl. Akad. Nauk SSSR, 150:275 (1963).
8. J. E. Geusic, W. B. Bridges, and J. I. Pankove, Proc. IEEE, 58:1419 (1970).
9. C. E. Hurwitz, Appl. Phys. Lett., 9:116 (1966).
10. A. R. Calawa, J. O. Dimmock, T. C. Harman, and I. Melngailis, Phys. Rve. Lett. 23:7 (1969).
11. D. Akerman, P. G. Eliseev, A. Kaiper, M. A. Man'ko, and Z. Raab, Kvant. Elektron. (Mosc.), No. 1, 85 (1971).
12. L. V. Keldysh, Zh. Eksp. Teor. Fiz., 33:994 (1957); 34:962, 1138 (1958).
13. V. S. Vavilov and K. I. Britsyn, Fiz. Tverd. Tela, 2:1937 (1960).
14. V. S. Vavilov, Effects of Radiation on Semiconductors, Consultants Bureau, New York (1965).
15. R. J. Phelan, A. R. Calawa, R. H. Rediker, R. J. Keyes, and B. Lax, Appl. Phys. Lett., 3:143 (1963).

16.   I. I. Zasavitskii, B. N. Matsonashvili, and A. P. Shotov, Fiz. Tekh. Poluprovodn., 4:337 (1970).

17.   L. P. Zverev, V. P. Bykov, G. M. Min'kov, S. A. Negashev, S. S. Khomutova, and V. Ya. Shur, Proc. Ninth Intern. Conf. on Physics of Semiconductors, Moscow, 1968, Vol. 1, Nauka, Leningrad (1968), p. 559.

18.   J. Feinleib, S. Groves, W. Paul, and R. Zallen, Phys. Rev., 131:2070 (1963).

19.   J. M. Besson, W. Paul, and A. R. Calawa, Phys. Rev., 173:699 (1968).

20.   H. R. Wittmann, Rev. Sci. Instrum., 39:1382 (1968).

21.   T. C. Harman, A. R. Calawa, I. Melngailis, and J. O. Dimmock, Appl. Phys. Lett., 14:333 (1969).

22.   P. Norton, P. Chia, T. Braggins, and H. Levinstein, Appl. Phys. Lett., 18:158 (1971).

23.   C. K. N. Patel and E. D. Shaw, Phys. Rev. Lett., 24:451 (1970).

24.   A. Mooradian, S. R. J. Brueck, and F. A. Blum, Appl. Phys. Lett., 17:481 (1970).

25.   I. I. Zasavitskii, B. N. Matsonashvili, and A. P. Shotov, Kratk. Soobshch. Fiz., No. 8, 81 (1971); Proc. Second Vavilov Conf. on Nonlinear Optics, Novosibirsk, 1971 [in Russian]; in: Nonlinear Processes in Optics [in Russian], No. 2, Nauka, Moscow (1972), p. 285.

26.   E. D. Hinkley, Appl. Phys. Lett., 16:351 (1970).

27.   K. W. Nill, F. A. Blum, A. R. Calawa, and T. C. Harman, Appl. Phys. Lett., 19:79 (1971).

28.   Yu. A. Bykovskii, V. L. Velichanskii, I. G. Goncharov, V. A. Maslov, and V. V. Nikitin, Opt. Spektrosk., 30:508 (1971).

29.   I. I. Zasavitskii, B. N. Matsonashvili, and A. P. Shotov, Zh. Prikl Spektrosk., 15:349 (1971).

30.   Yu. I. Ravich, B. A. Efimov, and I. A. Smirnov, Semiconducting Lead Chalcogenides, Plenum Press, New York (1970).

31.   B. M. Askerov, Transport Phenomena in Semiconductors [in Russian], Nauka, Leningrad (1970), p. 175.

32.   E. O. Kane, J. Phys. Chem. Solids, 1:249 (1957).

33.   L. M. Roth, B. Lax, and S. Zwerdling, Phys. Rev., 114:90 (1959).

34.   W. Zawadzki, Phys. Lett., 4:190 (1963).

35.   M. H. Cohen and E. I. Blount, Philos. Mag., 5:115 (1960).

36.   Y. Yafet, Solid State Phys., 14:1 (1963).

37.   M. Cardona, K. L. Shaklee, and F. H. Pollak, Phys. Rev., 154:696 (1967).

38.   O. Madelung, Physics of III-V Compounds, Wiley, New York (1964).

39.   R. K. Willardson and A. C. Beer, Semiconductors and Semimetals, Vol. 3, Optical Properties of III-V Compounds, Academic Press, New York (1967).

40.   F. Herman, R. L. Kortum, I. B. Ortenburger, and J. P. Van Dyke, J. Phys. (Paris), Suppl., 29:C4-62 (1968).

41.   R. Dalven, Infrared Phys., 9:141 (1969).

42.   D. L. Mitchell, E. D. Palik, and J. N. Zemel, Proc. Seventh Intern. Conf. on Physics of Semiconductors, Paris, 1964, Vol. 1, Physics of Semiconductors, publ. by Dunod, Paris; Academic Press, New York (1964), p. 325.

43.   K. F. Cuff, M. R. Ellett, C. D. Kuglin, and L. R. Williams, Proc. Seventh Intern. Conf. on Physics of Semiconductors, Paris, 1964, Vol. 1, Physics of Semiconductors, publ. by Dunod, Paris; Academic Press, New York (1964), p. 677.

44.   W. Paul, J. Appl. Phys. Suppl., 32:2082 (1961).

45.   J. Bardeen and W. Shockley, Phys. Rev., 80:72 (1950).

46.   W. Paul, Les Propriétés Physiques des Solides Sous Pression (Proc. Conf. Grenoble, 1969), Editions CNRS, Paris (1970), p. 199.

47.   J. O. Dimmock, I. Melngailis, and A. J. Strauss, Phys. Rev. Lett., 16:1193 (1966).

48.   H. Y. Fan, Photon-Electron Interaction: Crystals without Fields, Springer Verlag, Berlin (1967).

49. Yu. M. Popov, Tr. Fiz. Inst. Akad. Nauk SSSR, 31:3 (1965).

50. G. E. Pikus, Fundamentals of the Theory of Semiconductor Devices [in Russian], Nauka, Moscow (1965), p. 390.

51. M. H. Pilkuhn, Phys. Status Solidi, 25:9 (1968).

52. N. Holonyak, Jr., D. R. Scifres, M. G. Craford, W. O. Groves, and D. L. Keune, Appl. Phys. Lett., 19:256 (1971).

53. W. E. Ahearn and J. W. Crowe, IEEE J. Quantum Electron., QE-2:597 (1966).

54. E. D. Hinkley and C. Freed, Phys. Rev. Lett., 23:277 (1969).

55. B. D. Kopylovskii, V. S. Bagaev, Yu. N. Berozashvili, V. S. Ivanov, A. P. Shotov, and A. N. Khvoshchev, Prib. Tekh. Éksp., No. 4, 167 (1964).

56. B. D. Kopylovskii and V. S. Ivanov, Prib. Tekh. Éksp., No. 4, 145 (1965).

57. H. G. Häfele, H. Wachernig, C. Irslinger, R. Grisar, and R. Nitsche, Phys. Status Solidi, 42:531 (1970).

58. I. I. Zasavitskii, B. N. Matsonashvili, and A. P. Shotov, Prib. Tekh. Éksp., No. 2, 219 (1970).

59. I. A. Baranov, Yu. F. Bychkov, V. R. Karsik, and G. B. Kurganov, Élektron. Tekh. Ser. 1, Élektron. SVCh, No. 10, 104 (1957).

60. D. B. Montgomery, Solenoid Magnet Design, Wiley, New York (1969).

61. A. B. Fradkov, Prib. Tekh. Éksp., No. 6, 215 (1965).

62. L. F. Vereshchagin and V. E. Ivanov, Prib. Tekh. Éksp., No. 4, 73 (1957).

63. L. F. Vereshchagin and A. I. Likhter, Dokl. Akad. Nauk SSSR, 103:701 (1955).

64. A. I. Likhter, Prib. Tekh. Éksp., No. 2, 127 (1960).

65. V. V. Nefedova and A. P. Minin, Prib. Tekh. Éksp., No. 2, 198 (1973).

66. I. I. Zasavitskii, A. I. Likhter, É. G. Pel', and A. P. Shotov, Prib. Tekh. Éksp., No. 2, 203 (1973).

67. D. Langer and D. M. Warschauer, Rev. Sci. Instrum., 32:32 (1961).

68. W. Paul, W. M. DeMeis, and J. M. Besson, Rev. Sci. Instrum., 39:928 (1968).

69. Sh. M. Kogan (ed.), Photoconductivity [in Russian], Nauka, Moscow (1967), p. 91.

70. R. J. Keyes and T. M. Quist, "Low level coherent and incoherent detection in the infrared," in: Semiconductors and Semimetals (ed. by R. K. Willardson and A. C. Beer), Vol. 5, Infrared Detectors, Academic Press, New York (1970), p. 321.

71. Yu. D. Mozzhorin and V. I. Stafeev, Physics of p–n Junctions and Semiconductor Devices, Consultants Bureau, New York (1971), p. 65.

72. N. P. Esina, N. V. Zotova, and D. N. Nasledov, Radiotekh. Elektron., 8:1603 (1963).

73. A. P. Koshcheev, Diploma Thesis, Moscow State University (1970).

74. I. Melngailis and R. H. Rediker, J. Appl. Phys., 37:899 (1966).

75. A. I. Ansel'm, Introduction to the Theory of Semiconductors [in Russian], Fizmatgiz, Moscow–Leningrad (1962), p. 317.

76. S. M. Ryvkin, Photoelectric Effects in Semiconductors, Consultants Bureau, New York (1964).

77. I. Melngailis, IEEE J. Quantum Electron., QE-1:104 (1965).

78. I. D. Anismova, Yu. D. Mozzhorin, and V. M. Yungerman, Physics of p–n Junctions [in Russian], Zinatne, Riga (1966), p. 305.

79. T. I. Galkina, N. A. Penin, and V. A. Rassushin, Fiz. Tekh. Poluprovodn., 1:230 (1967).

80. N. P. Esina, V. N. Zotova, and D. N. Nasledov, Fiz. Tekh. Poluprovodn., 3:1370 (1969).

81. I. D. Anisimova, N. A. Ivashneva, and Yu. D. Mozzhorin, Fiz. Tekh. Poluprovodn., 3:1680 (1969).

82. I. G. Austin, J. Electron. Control, 8:167 (1960).

83. F. L. Galeener, I. Melngailis, G. B. Wright, and R. H. Rediker, J. Appl. Phys., 36:1574 (1965).

84. R. J. Phelan, Jr. and R. H. Rediker, Proc. IEEE, 52:91 (1964).

85. V. I. Stafeev, Fiz. Tverd. Tela, 1:841 (1959).

86. N. Holonyak, Jr., Proc. IRE, 50:2421 (1962).

87. I. Melngailis and R. H. Rediker, Proc. IRE, 50:2428 (1962).

88. A. P. Shotov, S. P. Grishechkina, and R. A. Muminov, Fiz. Tverd. Tela, 8:2496 (1966).

89. I. Melngailis, R. J. Phelan, and R. H. Rediker, Appl. Phys. Lett., 5:99 (1964).

90. A. P. Shotov, S. P. Grishechkina, B. D. Kopylovskii, and R. A. Muminov, Fiz. Tverd. Tela, 8:1083 (1966).

91. A. P. Shotov, S. P. Grishechkina, and R. A. Muminov, Zh. Éksp. Teor. Fiz., 50:1525 (1966); 52:71 (1967).

92. B. D. Osipov, and A. N. Khvoshchev, Zh. Éksp. Teor. Fiz., 43:1179 (1962).

93. A. P. Shotov, I. I. Zasavitskii, B. N. Matsonashvili, and R. A. Muminov, Proc. Ninth Intern. Conf. on Physics of Semiconductors, Moscow, 1968, Vol. 2, publ. by Nauka, Leningrad (1968), p. 895.

94. A. P. Shotov, S. P. Grishechkina, and R. A. Muminov, ZhÉTF Pis'ma Red., 6:895 (1967).

95. A. G. Chynoweth and S. J. Buchsbaum, Phys. Today, No. 11, 26 (1965).

96. V. A. Kovarskii and I. I. Zasavitskii, in: Research on Semiconductors [in Russian], Kishinev (1968), p. 123.

97. G. Bemski, Phys. Rev. Lett., 4:62 (1960).

98. S. Zwerdling, W. H. Kleiner, and J. P. Theriault, J. Appl. Phys. Suppl., 32:2118 (1961).

99. V. L. Bonch-Bruevich, in: Solid State Physics (Progress Review) (ed. by S. V. Tyablikov), VINITI, Moscow (1965), p. 127.

100. J. F. Butler and A. R. Calawa, Physics of Quantum Electronics (Proc. Intern. Conf., San Juan, Puerto Rico, 1965), McGraw-Hill, New York (1966), p. 458.

101. I. I. Zasavitskii, B. N. Matsonashvili, and A. P. Shotov, Fiz. Tekh. Poluprovodn., 6:1288 (1972).

102. G. E. Smith, G. A. Baraff, and J. M. Rowell, Phys. Rev., 135:A1118 (1964).

103. P. J. Lin and L. Kleinman, Phys. Rev., 142:478 (1966).

104. S. Rabii, Phys. Rev., 167:801 (1968).

105. G. A. Baraff, Phys. Rev., 137:A842 (1965).

106. P. M. Platzman, Phys. Rev., 139:A379 (1965).

107. A. L. McWhorter, Physics of Quantum Electronics (Proc. Intern. Conf., San Juan, Puerto Rico, 1965), McGraw-Hill, New York (1966), p. 111.

108. P. A. Wolff, Phys. Rev. Lett., 16:225 (1966); IEEE J. Quantum Electron., QE-2:659 (1966).

109. Y. Yafet, Phys. Rev., 152:858 (1966).

110. C. K. N. Patel and E. D. Shaw, Phys. Rev. B, 3:1279 (1971).

111. P. A. Wolff, Proc. Ninth Intern. Conf. on Physics of Semiconductors, Moscow, 1968, Vol. 1, publ. by Nauka, Leningrad (1968), p. 194.

112. V. P. Makarov, Zh. Éksp. Teor. Fiz., 55:704 (1968).

113. B. S. Wherrett and P. G. Harper, Phys. Rev., 183:692 (1969).

114. A. Mooradian and G. B. Wright, Phys. Rev. Lett., 16:999 (1969).

115. A. Moooradian and A. L. McWhorter, Phys. Rev. Lett., 19:849 (1967).

116. B. Tell and R. J. Martin, Phys. Rev., 167:381 (1968).

117. C. K. N. Patel and R. E. Slusher, Phys. Rev., 167:413 (1968).

118. C. K. N. Patel and R. E. Slusher, Phys. Rev. Lett., 21:1563 (1968).

119. R. E. Slusher, C. K. N. Patel, and P. A. Fleury, Phys. Rev. Lett., 18:77 (1967).

120. C. K. N. Patel, Proc. Symp. on Modern Optics, Polytechnic Institute, Brooklyn, New York, 1967, publ. by Wiley, New York (1967), pp. 19–51.

121. W. D. Johnston Jr and I. P. Kaminow, Phys. Rev., 168:1045 (1968); 178:1528 (1969).

122. W. D. Johnston Jr, I. P. Kaminov, and J. G. Bergman, Jr., Appl. Phys. Lett., 13:190 (1968).

123. C. K. N. Patel and R. E. Slusher, Phys. Rev., 177:1200 (1969).

124. S. R. J. Brueck and A. Mooradian, Appl. Phys. Lett., 18:229 (1971).

125. S. R. J. Brueck and A. Mooradian, Phys. Rev. Lett., 28:161 (1972).

126. R. L. Allwood, R. B. Dennis, R. G. Mellish, S. D. Smith, B. S. Wherrett, and R. A. Wood, J. Phys. C, 4:126 (1971).

127. C. K. N. Patel, Appl. Phys. Lett., 18:274 (1971).

128. R. L. Allwood, S. D. Smith, S. D. Devine, R. A. Wood, and R. G. Mellish, J. Phys. C, 3:186 (1970).

129. R. L. Allwood, R. B. Dennis, S. D. Smith, B. S. Wherrett, and R. A. Wood, J. Phys. C, 4:163 (1971).

130. E. D. Shaw and C. K. N. Patel, Appl. Phys. Lett., 18:215 (1971).

131. R. L. Aggarwal, B. Lax, C. E. Chase, C. R. Pidgeon, D. Limbert, and F. Brown, Appl. Phys. Lett., 18:383 (1971).

132. C. K. N. Patel, Appl. Phys. Lett., 19:400 (1971).

133. C. K. N. Patel, Phys. Rev. Lett., 28:649 (1972).

134. C. K. N. Patel, E. D. Shaw, and R. J. Kerl, Phys. Rev. Lett., 25:8 (1970).

135. C. Irslinger, R. Grisar, H. Wachernig, H. G. Häfele, and S. D. Smith, Phys. Status Solidi b, 48:797 (1971).

136. R. G. Mellish, R. B. Dennis, and R. L. Allwood, Opt. Commun., 4:249 (1971).

137. R. A. Wood, R. B. Dennis, and J. W. Smith, Opt. Commun., 4:383 (1972).

138. I. I. Zasavitskii, A. I. Likhter, É. G. Pel', and A. P. Shotov, Fiz. Tekh. Poluprovodn., 6:2206 (1972).

139. N. N. Shuikin, Thesis for Candidate's Degree, Lebedev Physics Institute, Moscow (1970).

140. A. A. Averkin, U. V. Ilisavski, and A. R. Regel, Proc. Sixth Intern. Conf. on Physics of Semiconductors, Exeter, England, 1962, publ. by The Institute of Physics, London (1962), p. 690.

141. G. Martinez, I. Chambouleyron, J. M. Besson, and M. Balkanski, Les Proprietes Physiques des Solides Sous Pression (Proc. Conf., Grenoble, 1969), Editions CNRS, Paris (1970), p. 241.

142. V. S. Bagaev, Yu. N. Berozashvili, V. S. Ivanov, B. D. Kopylovskii, and Yu. N. Korolev, Prib. Tekh. Éksp., No. 4, 185 (1966).

143. V. A. Ageikin, I. I. Zasavitskii, V. G. Koloshnikov, A. I. Likhter, É. G. Pel', and A. P. Shotov, Opt. Spektrosk., 36:808 (1974).

144. W. Kaiser, P. H. Kech, and C. F. Lange, Phys. Rev., 101:1264 (1956).

INVESTIGATION OF THE COLLECTIVE PROPERTIES
OF EXCITONS IN GERMANIUM BY LONG-WAVELENGTH
INFRARED SPECTROSCOPY METHODS*

V. A. Zayats

An investigation was made of the long-wavelength infrared transmission and luminescence spectra of germanium corresponding to the interband optical excitation at low temperatures. A new absorption and luminescence resonance with a maximum at $h\nu \approx 9 \times 10^{-3}$ eV was observed at $T \leq 2°K$. This resonance was interpreted by postulating that excitons in germanium condensed into electron–hole drops. The dipole absorption model was used to determine the plasma frequency and the particle density of these drops, as well as several other parameters. Photoionization and excitation spectra of free excitons in germanium illuminated with submillimeter radiation were determined at temperatures of 3-6°K. The binding energy in the ground state and the photoionization cross section of indirect excitons were determined.

INTRODUCTION

The Coulomb attraction between electrons and holes in a semiconductor crystal produces bound states of these two particles, in a manner similar to the Coulomb binding of an electron and a proton in the hydrogen atom. These bound states of electrons and holes, capable of moving across a crystal, are known as excitons [1, 2].

Excitons in semiconductor crystals are characterized by small (compared with atomic) binding energies and macroscopically large bound-state radii, because the Coulomb interaction in such crystals is weakened considerably by the high permittivity ($\varkappa \gtrsim 10$) and the effective masses of electrons and holes are relatively small ($m_{e,h} \sim 0.1 m_0$, where m_0 is the mass of a free electron). For example, in the case of an indirect exciton in germanium, the binding energy in the ground state is approximately 4×10^{-3} eV, and the corresponding Bohr radius is about 140 Å. Consequently, an exciton in a crystal can be regarded, in the first approximation, as a quasiatom in vacuum and the actual atomic structure of a semiconductor can be allowed for by the parameters m_e, m_h, and $\varkappa$.

In view of these length and energy scales in exciton systems, the interaction between excitons may be very strong even at relatively low exciton concentrations and the collective effects, reflecting the nature of the interexciton interaction, are unavoidable. The problem of collective properties of excitons was first encountered in theoretical consideration of the possibility of the Bose–Einstein condensation of excitons and of the associated superfluidity and superconductivity phenomena [3-5]. It has been shown that if the repulsive forces pre-

*Thesis submitted for the degree of Candidate of Physicomathematical Sciences, defended in 1972 at the P. N. Lebedev Physics Institute, Academy of Sciences of the USSR, Moscow.

dominate in the interaction between excitons, we may expect their Bose—Einstein condensation, but if the attractive forces are stronger, we may expect formation of exciton molecules (biexcitons) [6, 7].

Recently, L. V. Keldysh predicted theoretically a different behavior of a system of excitons when their concentration is high [8-10]. A qualitative proof of the predominance of the attractive interaction between excitons was followed by a hypothesis that when the concentration of excitons is increased, a literal condensation ("liquefaction") should occur. This transition in exciton systems should have many of the characteristic features of phase transitions of the first kind and the resultant regions of a dense electron—hole phase should have metallic conduction, i.e., they should resemble a liquid metal.

For some time this hypothesis of the liquefaction of excitons remained largely hypothetical because it was practically impossible to justify theoretically the existence of a system of particles with a concentration $na_{ex}^3 \sim 1$ (a_{ex} is the Bohr exciton radius). This hypothesis stimulated experimental investigations of high-concentration exciton systems in semiconductors, which were carried out recently using different experimental methods. These investigations established several points, which supported the idea of the existence of a condensed electron—hole phase in semiconductors.

The present paper is one of these first investigations and it reports an experimental study of the far-infrared absorption and luminescence spectra of excited germanium. In most semiconductors the binding energy of excitons corresponds to the far-infrared part of the spectrum, i.e., in studies of the collective exciton effects in semiconductors this part of the spectrum is analogous to the visible and near-infrared regions used in metal-optics investigations.

The idea of a possible condensation of excitons follows from an analysis of the properties of exciton systems based on the simple representation of excitons as hydrogen-like atoms. In fact, because of the complex energy structure of the majority of semiconductors, the structure of an exciton may differ very considerably from the simple hydrogen model. Studies of the real energy spectra of excitons are, therefore, not only of intrinsic interest but should help in the understanding of various exciton phenomena which occur in semiconductors, and these include collective effects.

It is difficult to carry out theoretical calculations of the energy spectra of excitons allowing for the real energy-band structure and such calculations give usually results of limited usefulness. This is why many experimental investigations have been carried out employing various methods, including optical and magnetooptic absorption, photoluminescence, photoconductivity, differential spectra, etc. (see, for example, review [11]). However, all these methods are based on investigations of the optical phenomena in semiconductors in the same spectral range near the fundamental absorption edge. This makes it difficult to obtain information on the energy structure of excitons, such as the excited-state spectrum. This is particularly true of the "indirect" excitons in crystals such as germanium and silicon. Investigations of the optical properties of excitons in the far infrared, i.e., in the region of their ionization, provide the most direct experimental method for determining the energy spectra of excitons.

In the case of a different fine electron system, represented by group III and V impurities in germanium and silicon, investigations in an analogous range of energies have yielded not only detailed information on the energy structure of impurity centers [12, 13] but have revealed experimentally various phenomena associated with the exchange interaction between neighboring impurity centers present in a sufficiently high concentration [14, 15].

The present paper describes an experimental investigation of the energy spectra and collective properties of excitons based on the long-wavelength infrared spectroscopy. This method was applied for the first time to the study of exciton phenomena in semiconductors.

All the measurements were carried out on germanium crystals in which high steady-state indirect exciton concentrations could easily be established because of the long exciton lifetime. Another important point was the availability of the results of thorough investigations of the long-wavelength infrared spectra of intrinsic and extrinsic germanium under equilibrium conditions (in the absence of excitation).

CHAPTER I

ENERGY SPECTRA AND COLLECTIVE PROPERTIES OF EXCITONS IN SEMICONDUCTORS

PART 1. ENERGY SPECTRUM OF EXCITONS

§1. Theoretical Calculations

Allowance for the Coulomb attraction between negatively charged conduction-band electrons and valence-band holes, which behave as positively charged particles, produces bound electron–hole states in semiconductors and these are known as excitons. Since the Coulomb interaction in a medium with a high permittivity $\varkappa$ (in the case of semiconductors, $\varkappa \sim 10$) is much weaker than in free space, the binding energy of an exciton $\mathscr{E}_{ex}$ is relatively small and its radius a_{ex} is relatively large. In any case, we find that

$$\mathscr{E}_{ex} \ll E_g, \ a_{ex} \gg d \tag{1}$$

(E_g is the forbidden-band width and d is the lattice constant). The small effective masses of electrons and holes only strengthen the inequalities of Eq.(1). It follows that the energy spectrum of excitons can be calculated using the effective mass approximation in which an exciton can be treated as a two-body system.

In the case of the simple model of spherical nondegenerate bands, considered in [2, 16, 17] the Schrödinger equation for such a system has the form

$$\left(-\frac{\hbar^2 \nabla_e^2}{m_e} - \frac{\hbar^2 \nabla_h^2}{m_h} - \frac{e^2}{\varkappa r}\right)\Psi = \mathscr{E}\Psi, \tag{2}$$

where $r = |\mathbf{r}| = |\mathbf{r}_e - \mathbf{r}_h|$; $\mathbf{r}_e$ and $\mathbf{r}_h$ are the radius vectors of an electron and a hole; the Hamiltonian of the system is identical with the Hamiltonian of the hydrogen atom with an effective nuclear charge $e/\varkappa$ and a reduced mass $m_r = [(1/m_e) + (1/m_h)]^{-1}$. The solution of Eq.(2) is the complete wave function of the system:

$$\Psi_{nlm, \mathbf{k}} = e^{i\mathbf{k}\rho} \cdot F_{nlm}(\mathbf{r}), \tag{3}$$

where ρ is the radius vector of the center of mass of an electron–hole pair; F_{nlm} are the suitably modified wave functions of a hydrogen-like atom; n, l, and m are the quantum numbers. The energy levels of the bound states corresponding to each value of the wave vector k are measured from the total dissociation state when an electron and a hole are an infinite distance apart:

$$\mathscr{E}_n(\mathbf{k}) = -\frac{m_r e^4}{2\hbar^2 \varkappa^2 n^2} + \frac{\hbar^2 k^2}{2(m_e + m_h)}, \tag{4}$$

where n = 1, 2, 3, ... is the principal quantum number. The first term in Eq. (4) gives the energy levels (orbitals) of a hydrogen-like atom $\mathcal{E}_n(0) \propto 1/n^2$, associated with the relative motion of an electron and a hole in an exciton around a common center of mass. The second term gives the kinetic energy of motion of an exciton as a whole, whose mass is $m_{ex} = m_e + m_h$ and whose quasimomentum is $p_{ex} = \hbar k_{ex} = \hbar(k_e + k_h)$. We can see that if allowance is made for the motion of the center of mass of an electron–hole pair, the exciton energy levels transform into energy bands and, as in the case of electrons, it is then sufficient to consider the dispersion of $\mathcal{E}_n(k)$ within the first Brillouin zone. The minimum of the exciton energy in a band, i.e., the rest energy of an exciton, in a direct-gap semiconductor with extrema of both bands at $k_e^0 = k_h^0 = 0$ is also located at $k_{ex}^0 = 0$. Such excitons are known as direct. In the case of indirect-gap semiconductors we have $k_e^0 \neq 0$ (or $k_h^0 \neq 0$) and Eqs.(3) and (4) remain the same but now the exciton energy minimum corresponds to $k_{ex}^0 = k_e^0 + k_h^0 \neq 0$. Such excitons are known as indirect.

A hydrogen-like series of absorption lines of frequencies ν_n described by an expression $h\nu_n = E_g - \mathcal{E}_n(0)$, was first observed in cuprous oxide crystals [18], but this semiconductor has since been found to be almost the only material with a sharp hydrogen-like exciton spectrum. In the case of cuprous oxide we have that rare situation that there are two band edges with spherical energy surfaces at $k = 0$.

A simple band model with some corrections to Eq.(4) due to degeneracy of the valence-band edge, found by the perturbation theory, yields the binding energy of indirect excitons in germanium $\mathcal{E}_{ex} \approx 5 \times 10^{-3}$ eV [17] and the corresponding Bohr radius

$$a_{ex} = \frac{\hbar^2 \varkappa}{e^2 m_r} = \frac{e^2}{2 \varkappa \mathcal{E}_{ex}} \simeq 90\,\text{Å}. \tag{5}$$

It is worth pointing out that in this connection the inequalities of Eq.(1) are satisfied quite well, so that the effective mass approximation can be applied to germanium.

In the case of real band structures of such semiconductors as germanium and silicon the theoretical discussion of electron states is much more complex [17]. The problem of description of shallow bound states subject to allowance for the real band structure of germanium and silicon is considered for III and V group impurities in [19]. The similarity of the theoretical models of shallow impurities and excitons makes it possible to apply the ideas developed in [19] to shallow centers and, more specifically, to excitons in the same semiconductors [20]. Both impurity and exciton states are formed in a similar manner from electron states of the conduction and valence bands. However, whereas in the case of shallow centers it is sufficient to allow for the contribution of just one nearest band (which is the conduction band for donors and the valence band for acceptors), the exciton wave functions are formed from electron states in both bands. The theory of shallow donors and acceptors in germanium and silicon [19] and calculation methods for the determination of the energy spectrum of shallow centers in such systems [21] are combined in [20] in a calculation of the energy levels of the ground states of excitons in germanium and silicon.

The valence-band states in germanium corresponding to low values of k are described in [20] using a (4×4)-matrix effective-mass Hamiltonian for holes $H^{(v)}(k)$, obtained from the complete (6×6) matrix of the valence band [19] ignoring the contribution made to the exciton wave functions by the states in the band split off by the spin–orbit interaction. The large, compared with the exciton binding energy, spin–orbit splitting in germanium (0.28 eV) justifies this approach. The effective-mass Hamiltonian for electrons in the case of indirect excitons in germanium $H^{(c)}(k - k_e^0)$ is constructed from the conduction-band states at four equivalent minima along the [111] direction near $k = k_e^0$. The constant-energy surfaces are assumed to be ellipsoids of revolution strongly elongated along the [111] direction and characterized by longitudinal and transverse components of the electron effective mass. The exciton Hamiltonian is assumed to be

$$H_{ij} = \left[H^{(e)} (-i\nabla) - \frac{e^2}{\varkappa r} \right] \delta_{ij} - H^{(v)}_{ij} (-i\Delta). \tag{6}$$

Here, the index i = 1, 2, 3, or 4 indicates summation over four degenerate (allowing for spin) valence bands at $\mathbf{k} = 0$. In the case of a degenerate valence band the functions $F^{(i)}_{n, \mathbf{k}^0_e} (\mathbf{r})$, describing the relative motion of an electron and a hole should satisfy the following system of coupled differential equations:

$$\sum_j H_{ij} F^{(j)}_{n, \mathbf{k}^0_e} (\mathbf{r}) = \mathcal{E}_{n, \mathbf{k}^0_e} F^{(i)}_{n, \mathbf{k}^0_e} (\mathbf{r}). \tag{7}$$

The Hamiltonian H_{ij} of the type given by Eq.(6) does not give an exact solution of the system (7) for $F^{(i)}_{n, \mathbf{k}^0_e} (\mathbf{r})$. The variational calculation method developed in [21] for shallow acceptor centers in germanium was applied in [20] to the ground-state energies of direct and indirect excitons in germanium. Similar calculations were also made for indirect excitons in silicon. The test functions for $F^{(i)}$ were either s-type, or s- and d-type. Allowance for the d-type states was found to be necessary in the calculation of the ground-state energy of indirect excitons in germanium. In this case the calculations yielded the energies of two levels corresponding to the splitting of the ground state of indirect excitons.

The existence of the splitting of indirect exciton states in germanium may be deduced from symmetry considerations [22] and it can be explained qualitatively as follows. The degeneracy of the valence-band edge of germanium leads to a degeneracy of the exciton energy levels. In the case of indirect excitons in germanium this degeneracy is partly lifted by the presence of several equivalent minima of the conduction band with an anisotropic electron mass, which results in the splitting of the electron states into two different levels. In the case of direct excitons in germanium the degeneracy of the energy levels is retained, like the degeneracy of the acceptor levels [22], because in this case electrons are located in the only minimum of the conduction band at $\mathbf{k} = 0$, which is of very simple type.

The results of calculations of the ground-state energy levels of direct and indirect excitons in germanium, obtained using test functions of the s- and (s + d)-type, are given in Table 1.

It is clear from this table that the use of the (s + d)-type test functions yields two ground-state levels separated by $\Delta_{ex} = \mathcal{E}_{ex_1} - \mathcal{E}_{ex_2} = 0.59$ meV and the average of these two levels is $(\mathcal{E}_{ex_1} + \mathcal{E}_{ex_2})/2 = 3.17$ meV.

The same ideas, developed in [19] for shallow impurity centers in germanium, were used in [23] to calculate the energies of the two lowest levels of indirect excitons in germanium subject to certain simplifying assumptions. For example, the (4 × 4)-matrix effective-mass Hamiltonian of holes was replaced with the diagonal elements because the nondiagonal elements were found to be small when the z axis was selected along the [111] preferred direction in

TABLE 1

Type of exciton	$F^{(i)}$	ε, meV
Direct	s	1.27
	$(s+d)$	1.38
Indirect	s	2.44
	$(s+d)$	3.47 $(\mathcal{E}_{ex_1})$
		2.88 $(\mathcal{E}_{ex_2})$

germanium. Moreover, when the axes were selected in this way, the diagonal terms were pairwise equal and they corresponded to two ellipsoids, so that two effective masses with appropriate longitudinal and transverse components could be introduced. These components were determined using the parameters γ of the valence band suggested in [24] and the experimental data on cyclotron resonance. Combining in turn the two ellipsoids obtained for the valence band with the constant-energy ellipsoid of the conduction-band electrons, it was found that the transformation to the coordinates of the center of mass of an exciton gave two reduced ellipsoids describing thre relative motion of an electron and a hole in an exciton by two sets of transverse and longitudinal components of the reduced effective mass. Thus, the problem of finding exciton levels was divided into two stages, each of which was fully analogous to the calculation of the energy spectrum of shallow donors of germanium [19].

Applying the variational principle and taking the test function in the same form as in [19], the ground-state energy was calculated in both cases in [23] and in this way the following two levels were obtained for the split ground state of indirect excitons in germanium: $\mathscr{E}_{ex_1} = 3.3$ meV, $\mathscr{E}_{ex_2} = 2.5$ meV, $(\mathscr{E}_{ex_1} + \mathscr{E}_{ex_2})/2 = 2.9$ meV, and $\Delta_{ex} = 0.8$ meV.

It should be noted that in this calculation method the splitting of the ground state into two levels appeared in a natural manner so that these levels did not belong to the same hydrogen-like series like the ground (n = 1) and first excited (n = 2) levels.

It was held in [23] that the inclusion of the diagonal terms could increase the splitting Δ_{ex} and the average energy of the two exciton levels in the ground state. A more accurate calculation [20] did indeed give a higher value for the average energy of the two levels, but a slightly smaller splitting (Table 1).

The spectrum of excited states of an exciton was not calculated in [20] or [23] because of considerable mathematical difficulties.

§2. Experimental Results

The published experimental data on the energy levels of excitons in germanium and silicon were obtained from careful studies of the fine structure of the fundamental absorption edge of these compounds and were restricted to the binding energies of excitons in the ground state.

Transitions to exciton states contribute to the long-wavelength part of the fundamental absorption edge either in the form of an absorption line (direct excitons) or in the form of steps in the absorption curve (indirect excitons) [25]. In the latter case the exciton absorption can be separated from the overall absorption curve because of the different dependence of the absorption coefficient on the photon energy in the interband and exciton transition regions [17, 26]. Obviously, in such studies the precision of the determination of the energy of exciton levels cannot be high.

The method described above was used in [25] to find the binding energy of indirect excitons in germanium and the initial value was $\mathscr{E}_{ex} \approx 5$ meV. An improvement of the method and a more careful analysis of the experimental results made it possible to refine these data. In the latest of this series of papers [25] the energies of the two levels of indirect excitons in germanium were quoted as 2.7 and 1.7 meV. The structure of the fundamental absorption edge of germanium was investigated in [23] using magnetic fields (magnetoabsorption). In this case it was somewhat easier to separate the exciton absorption from the general curve by using the different dependences of the energies of the interband (occurring between Landau levels) and exciton transitions on the applied magnetic field [23]. The binding energy of excitons was determined by extrapolation of these dependences to zero magnetic field and a value of 2.5 meV was obtained for indirect excitons in germanium. Theoretical calculations carried out by the same authors (see §1) showed that the ground state of indirect excitons in germanium should be split into two levels. A more careful analysis of the experimental spectra by

TABLE 2

Direct excitons	Indirect excitons				Reference
$\mathscr{E}$, meV	$\mathscr{E}_{ex_1}$, meV	$\mathscr{E}_{ex_2}$, meV	$\dfrac{\mathscr{E}_{ex_1}+\mathscr{E}_{ex_2}}{2}$	Δ_{ex}, meV	
Theoretical results					
1.4	3.5	2.9	3.2	0.6	[20]
1.7	3.3	2.5	2.9	0.8	[23]
Experimental results					
1.1±0.1	3.3±0.4	2.3±0.4	2.8±0.4	1.0±0.1	[23]
2.5±0.5	2.7±0.4	1.7±0.4	2.2±0.4	1.0±0.1	[25]
—	3.6±0.3	2.8±0.3	3.2±0.3	0.8±0.1	[28]

differentiation [23] revealed this splitting and gave the following energies for the two exciton levels: 3.3 and 2.3 meV.

Recently, a more sensitive differential method was developed for the recording of the radiation transmitted by a crystal and this made it possible to determine more accurately the binding energy of indirect excitons in silicon [27] and germanium [28]. In the case of germanium it was found that $\mathscr{E}_{ex} = 3.6 \pm 0.3$ meV [28]. Moreover, the binding energy of indirect excitons could be deduced from the position of the recombination radiation (luminescence) line of indirect excitons [29, 30] provided the forbidden-band width at a given temperature and the energy of phonons participating in the optical transitions were known sufficiently accurately.

The theoretically calculated values of the energies of exciton levels in germanium and silicon are in reasonable agreement with the values found in different experiments (Table 2). The differences may be attributed to the low precision of the experimental methods rather than to deviations of the theoretical models from real systems or the approximate nature of the calculations. The same theoretical model gives a very good agreement between calculations and experiment in the case of various shallow impurities in germanium [19, 21]. Some discrepancies are found in respect of the ground-state energies of impurity centers but this is due to the violation of the conditions of validity of the effective-mass approximation at short distances from an impurity atom. We may expect these discrepancies to be even less in the case of excitons because the Bohr radius of an exciton is several times greater than the Bohr radius of an impurity center and, moreover, in contrast to impurity atoms, excitons do not give rise to local distortions in the host lattice.

PART 2. COLLECTIVE PROPERTIES OF EXCITON SYSTEMS

In Part 1 of the present chapter, we have considered the energy spectrum of excitons on the assumption that there is no interaction between them. In particular, we have shown that a typical binding energy of an exciton in a semiconductor crystal is $\mathscr{E}_{ex} \sim 10^{-2}$ eV, which is approximately three orders of magnitude less than a typical atomic energy, and that the exciton-state radius $a_{ex} \sim 10^{-6}$ cm is more than an order of magnitude larger than atomic distances in a crystal. This very high value of a_{ex} makes the product $(n_{ex} a_{ex}^3)^{1/3}$ approach unity for $n_{ex} \sim 10^{15}$-10^{16} cm^{-3}, i.e., in this situation the average distance between the excitons becomes comparable with the distance between the particles in the excitons. Clearly, under such conditions, the properties of the whole exciton system are largely governed by the exciton–exciton interaction and collective effects reflecting this interaction must necessarily appear in the system. No special experimental difficulties are encountered in generating excitons in the concentration mentioned above or even in higher concentrations if pure semiconductor crystals are used.

At low temperatures ($kT \ll \mathcal{E}_{ex}$) and very low exciton concentrations, when the condition $n_{ex} a_{ex}^3 \ll 1$ is obeyed by a large margin, the interaction of excitons with one another and with the crystal lattice can be regarded as weak and an exciton system of this kind can be regarded as a special type of an ideal gas of quasiatoms. The internal structure of an exciton is governed, as in the case of a normal atom (such as the hydrogen atom), by the Coulomb interaction between the negative and positive particles (see Part 1), except that the heavy proton in an atom is replaced by a hole with a mass of the order of the electron mass. The actual atomic structure of a semiconductor may be important, because of the macroscopic dimensions of excitons in crystals, only to the extent that it determines the parameters $m_{e,h}$ and $\varkappa$. The distortion of the crystal structure itself by the presence of excitons even in high concentrations is negligible. Thus, in the first approximation, we may assume that excitons in a crystal behave as quasiatoms in vacuum and the behavior of a system of excitons (or nonequilibrium electrons and holes) at higher concentrations should be largely similar to the behavior of normal atoms (or electrons and ions) when their concentration is raised. This analogy provides essentially the theoretical basis of the properties of excitons in the most interesting range of their concentrations $n a_{ex}^3 \sim 1$ because it is practically impossible to calculate exactly the properties and behavior of a system of particles with the Coulomb interaction in this range of concentrations.

§1. Theoretical Representations

At the lowest concentrations, a system of excitons behaves like an atomic gas. Exciton molecules (biexcitons) may appear in such a system at higher concentrations [6, 7]. We shall discuss the possibility of the existence of biexcitons at different concentrations from the point of view of the biexciton stability. When the concentration of an ordinary gas is increased sufficiently, the attraction between its atoms or molecules results in liquefaction, i.e., in the formation of a dense condensed phase in which the internal forces hold all the particles at distances of the atomic order. A qualitative conclusion that the interaction between excitons is predominantly attractive is used in [8] to show that a condensation-like transition should occur in an exciton system when an appropriate concentration is reached. The density of an exciton condensate n_0 and the temperature range of its existence should be governed by the conditions $n_0 a_{ex}^3 \sim 1$ and $kT_{cr} \ll \mathcal{E}_{ex}$, allowing for the appropriate scale of length and energies in an exciton system (in the case of germanium, we have $n_0 \sim 10^{17}$ cm^{-3} and $T_{cr} \ll 40°K$).

At higher concentrations in the range $n a_{ex}^3 \gg 1$, excitons can no longer exist and dissociate into almost free electrons and holes [31]. At these concentrations the Coulomb potential is screened over distances much shorter than the Bohr exciton radius and excitons cannot appear. Electrons and holes are then strongly degenerate. The kinetic (Fermi) energy of each of these particles is proportional to $n^{2/3}$ and exceeds considerably the potential energy of their interaction (which increases proportionally to $n^{1/3}$). Naturally, under these conditions, the bound exciton states cannot exist.

We shall now consider briefly the energy stability of biexcitons. The hypothesis of the formation of exciton molecules in nonmetallic crystals was put forward in [6, 7]. The possibility of the existence of a positronium molecule was demonstrated earlier in [32]. From the point of view of stability, the situation in a positronium molecule is most unfavorable because the electron and positron masses are equal. An estimate of the dissociation energy of the positronium molecule gives a very small value of the order of 2% of the binding energy of positronium (6.8 eV). Experiments have failed to confirm the existence of positronium molecules.

As in the case of positronium, the absence of heavy particles (such as the proton in the hydrogen molecule) from a biexciton results in a large (of the order of the exciton radius) amplitude of zero-point vibrations and a small (much smaller than the binding energy of an exciton) dissociation energy. Even when the mass ratio is $m_h/m_e \lesssim 10$ (this case approaches the situation in the hydrogen molecule), an estimate of the dissociation energy of an exciton molecule gives [33] a value not exceeding $(1/6)\, \mathcal{E}_{ex}$. It follows clearly from this discussion that the conditions for the appearance of undissociated exciton molecules would be difficult to achieve, particularly in the case of semiconductors when the electron and hole masses are similar.

The nature of the transition of an exciton gas to a dense condensed phase and its principal properties were first considered by L. V. Keldysh [9, 10]. A rigorous theoretical study of these questions met with considerable difficulties but the main properties of the exciton condensate and formation process were predicted from general considerations.

The absence of heavy particles in the condensed exciton phase means that a spatial order of the type encountered in crystals is impossible at any temperature because the amplitude of the zero-point vibrations of the particles should be of the order of a_{ex}, i.e., the average distance between the particles. In this respect, an exciton system is similar to liquid helium, which does not solidify at normal pressures right down to temperatures close to 0°K because of large zero-point vibrations.

Similar considerations indicate that it is unlikely that the condensed phase may consist of biexcitons. The large zero-point vibrations and the low binding energy of biexcitons should result in a strong interaction of each particle with all the nearest neighbors, strong electron exchange, and, consequently, collectivization of all electrons and holes. Therefore, the dense condensed exciton phase should resemble a liquid metal with an atomic structure rather than an insulating molecular liquid such as liquefied hydrogen.

The behavior of an exciton system during condensation and in the critical state preceding condensation can be considered by analogy with the behavior, under similar conditions, of alkali metal atoms or hydrogen. In fact, an exciton regarded as a quasiatom with one s valence electron resembles atoms in these substances. However, alkali metals and hydrogen behave quite differently during condensation. An alkali metal vapor at temperatures close to the critical value ($T_{cr} \sim 2000\text{-}3000°K$) consists largely of separate atoms because the dissociation energy of alkali metal molecules is low ($\lesssim 1$ eV). Condensation of an alkali metal vapor produces an atomic metal liquid. The dissociation energy of the hydrogen molecule is considerably higher (≈ 4.5 eV) and the low-temperature ($\approx 20°K$) liquefaction of the molecular gas produces a molecular liquid with an insulator spectrum. Hydrogen can be transformed into an atomic metallic phase only by the application of very high pressures.

Since the binding energy of a biexciton is relatively small, particularly in the case of similar values of the effective masses of an electron and a hole, we may expect the behavior of the exciton system in the critical state and during condensation to be similar to the behavior of alkali metal vapors. Thus, it is unlikely that biexcitons exist in the condensed phase or in the state preceding condensation at $T \approx T_{cr}$. We can use the analogy with liquid metals to estimate the critical transition temperature from the relationship $kT_{cr} \approx 0.1\, \mathcal{E}_{ex}$. In the case of germanium, we have $\mathcal{E}_{ex} \approx 4$ meV and $T_{cr} \sim 5°K$.

The transition from a gas of free excitons to a metallic electron–hole liquid should have many of the characteristic features of a phase transition of the first kind. In particular, when the average exciton concentration in a sample n reaches a certain value n_T, which depends strongly on temperature, a system should split into two phases: in an exciton gas of relatively low concentration, there should be regions of the condensed phase with a high concentration n_0 (liquid drops). At temperatures much lower than the critical value, such condensed drops may appear even for $n \ll n_0$; the liquid phase, in this case, occupies a small proportion of the total volume n/n_0. When the exciton concentration is increased, the volume of the liquid phase grows but its concentration n_0, governed by the internal interaction forces, remains constant until the liquid occupies the whole sample. The exciton concentration at which liquid drops appear is governed by the difference between the energies per pair of particles in the gaseous and condensed phases and it may be estimated from thermodynamic relationships. The formation of nuclei of the new phase should be stimulated, as in all other real condensation processes, by the presence of condensation centers which can be various inhomogeneities in crystals such as local stresses, dislocations, pile-ups of defects or impurities, etc. The critical concentration n_{cr} at which the drop evaporation temperature reaches its maximum value T_{cr} should be governed, by analogy with alkali metals, by the relationship $n_{cr} a_{ex}^3 \sim 10^{-1}$-$10^{-2}$. In the case of germanium, we have $n_{cr} \sim 10^{15}$-10^{16} cm^{-3}.

The distribution of particles between the gaseous and condensed phases cannot be governed simply by thermodynamic relationships. Since the excitons have a finite lifetime and continuous excitation is needed to maintain a given exciton concentration, it follows that kinetic or transport equations may be used to describe the balance of particles [10, 34]. An analysis of these equations shows that the condensed phase appears when the pair-generation rate is $G \gtrsim G_T = n_T/\tau_{ex}$ (τ_{ex} is the exciton lifetime in the gaseous phase), i.e., there is an excitation threshold below which this phase does not appear.

The foregoing considerations are based on the assumption that the binding energy per pair particles in the liquid phase is higher than in a biexciton and the transformation of excitons at higher concentrations to the liquid state is more likely from the energy point of view than the formation of biexcitons. This conclusion, drawn from qualitative considerations, may be proved or disproved only experimentally. It is possible that opposite situations may exist in different semiconductors. The condensation process should also occur in a biexciton system when the necessary concentration is reached because the dissociation energy of biexcitons decreases with rising concentration [10, 31]. Obviously, the concentration and temperature needed for this transition should differ from those estimated for excitons. Moreover, the properties of the exciton and biexciton systems should be different.

We shall conclude by pointing out that different coherent states are, in principle, possible in a nonequilibrium system of electrons and holes in semiconductors. These states, including the possibility of the Bose condensation of excitons and biexcitons and the formation of a superfluid or superconducting state in a liquid phase, are discussed in [3-5, 9, 10, 31], but these subjects are outside the scope of our experimental investigation of the collective properties of excitons in germanium.

§2. Discussion of Experimental Results

Exciton systems have not yet been investigated at high concentrations that necessarily lead to collective effects.

In 1966, Haynes discovered a new line in the recombination radiation spectrum of pure silicon at helium temperatures [35]. The new broad line was displaced relative to the indirect exciton line by 15 meV in the direction of longer wavelengths and its intensity at moderate pair generation rates G increased proportionally to G^2. These two observations led Haynes to the conclusion that the new line was due to the recombination radiation of exciton molecules. Haynes proposed a radiative recombination mechanism of complexes in which one of the excitons forming a molecule would recombine, giving up part of its energy to another exciton and causing the latter to dissociate into a free electron and a free hole. This mechanism explained satisfactorily the energy position and profile of the new line on the assumption that the binding energy of indirect excitons in silicon was 8 meV. Subsequent refinement of the binding energy of excitons in silicon (≈ 14 meV) [27, 36] indicated that the biexciton recombination mechanism proposed by Haynes [35] failed to explain at least the profile of the short-wavelength edge of the new line.

Metallic conduction in a high-concentration exciton system was discovered in 1967 by Asnin, Rogachev, and Ryvkin [37]. They studied the low-temperature photoconductivity of thin germanium samples and the study was carried out in a wide range of nonequilibrium carrier densities generated by optical excitation. At densities below 2×10^{15} cm^{-3}, there was practically no conductivity, which was attributed to the binding of free electrons and holes into excitons. The appearance and strong rise of the conductivity at $(2-3) \times 10^{15}$ cm^{-3} was attributed to the transition of an insulating exciton gas to a state with a metallic conduction (Mott transition). After a jump in this density range, the conductivity remained approximately constant. The authors concluded that both electrons and holes remained nondegenerate in this density range and the conductivity did not increase with density because of the fall in mobility. At a higher density ($n \approx 2 \times 10^{16}$ cm^{-3}), a second sudden jump in the conductivity was observed. It was attributed to the Fermi degeneracy of the carriers when the scattering of particles by one another was suppressed and the mobility rose with density. Thus, in order to explain the experimental results, it was assumed that the Mott transition occurred to a nondegenerate state, which was in conflict with the ideas about the Mott transition [38].

The first experimental investigations which provided evidence of possible condensation of excitons in germanium were reported in [39, 40].

Asnin and Rogachev [39] investigated the transmission of pure germanium at liquid helium temperatures in the direct-exciton range. The transmission was studied as a function of the concentration of electron–hole pairs excited in a sample. When the average (over a sample) concentration of these pairs was $\bar{n} \sim 10^{15}$-10^{16} cm^{-3}, the samples exhibited a bleaching effect and the form of the bleaching spectra was not affected up to $\bar{n} \approx 1 \times 10^{16}$ cm^{-3}, whereas the degree of bleaching rose rapidly (approximately in proportion to $\bar{n}^3$) with increasing excitation rate. It was concluded that these results indicated the formation of metallized regions (drops) composed of condensed excitons with a constant excitation-independent particle concentration n_0. The absorption in the direct-exciton state in this region should disappear because of the screening of the Coulomb interaction between electrons and holes by carriers in the metallic phase. The form of the bleaching spectra began to change at $\bar{n} \gtrsim 10^{16}$ cm^{-3}, when the condensed phase filled the whole sample and the concentration of the particles began to rise again. According to the results reported in [39], the equilibrium concentration in the liquid phase was $n_0 \approx 10^{16}$ cm^{-3}.

Observations of the condensed phase of nonequilibrium carriers in germanium were announced, about the same time as [39], in [34, 40]. In these two papers, Pokrovskii and Svistunova observed a new line (709.6 meV), similar to that reported by Haynes for silicon [35], in the photoluminescence spectrum of intrinsic germanium recorded at helium temperature.

The new broad line was shifted from the free-exciton line by 4.6 meV in the direction of lower energies: it appeared at $T \approx 2.6°K$ for $\bar{n} \sim 10^{14}$ cm^{-3} and its intensity rose rapidly when the temperature was lowered. The rise in the intensity of this line with the rate of excitation followed a cubic law in a wide range of concentrations but the rise became linear at the highest excitation rates. The quantum efficiency of the new luminescence line was close to unity (0.8-1.0) and the carrier lifetime in the condensate was 20 μsec. These parameters were, respectively, 10^{-2} and 1 μsec in the case of free excitons.

According to Pokrovskii and Svistunova, the new line was not due to the recombination of biexcitons but resulted from the radiative annihilation of nonequilibrium carriers forming an electron–hole condensate. They suggested a model of the condensed phase, which explained qualitatively the position and width of the luminescence line, as well as other properties such as the threshold nature of its appearance when the temperature was lowered or the concentration increased, nature of the dependence of the line intensity on the excitation rate, and high (compared with free excitons) values of the quantum efficiency of the luminescence and of the carrier lifetime in the condensate. This model was used to estimate the equilibrium concentration of the particles in the condensed phase, which was $n_0 \approx 2 \times 10^{17}$ cm^{-3}. The profile of the luminescence line of electron—hole drops in germanium, calculated later [41] for an equilibrium particle concentration of $n_0 = 3 \times 10^{17}$ cm^{-3}, was in good agreement with the experimental data.

An investigation of the behavior of a new luminescence line of germanium subjected to a uniaxial compression along the [111] axis was reported in [42] and along other crystallographic directions ([100] and [110]) in [43]. In contrast to the free-exciton luminescence line, whose position followed the reduction in the forbidden-band width, the position of the new line was practically constant up to a certain critical pressure P_{cr}. At pressures $P > P_{cr}$, the maximum of the new line followed the reduction in the forbidden-band width.

This behavior of the new line was explained qualitatively on the basis of a theory of electron—hole drops in deformed crystals. The change in the band structure of germanium under pressure caused the electron—hole drops to undergo a transition to a new equilibrium state with the lower equilibrium particle concentration and, consequently, with a lower binding energy per pair. Thus, the energy separation between the maxima of the new line and of the free-exciton line decreased to a certain minimum value at pressures $P < P_{cr}$ (the actual values depended on the direction of compression), but in the range $P > P_{cr}$ this separation remained practically constant.

A catastrophic fall in intensity of the new line due to inhomogeneous deformation was reported in the same papers [42, 43]. In this case, the intensity of the exciton line remained practically constant. This unusual experimental observation was explained by the hypothesis of the expulsion of drops by the deformation gradient from the part of the sample filled with excitons. Estimates given in [42] indicated that the electron—hole drops should have an extremely high mobility because of the very low effective mass density ($\sim 10^{-10}$ g/cm^3) and because the phonon scattering of the electrons and holes in the drops was suppressed by the Fermi degeneracy of the carriers. Inhomogeneous deformations (produced by a relatively small pressure gradient of ~ 100 kg/cm^2 per 1 cm) could accelerate electron—hole drops during their lifetime to velocities close to the velocity of sound.

Thus, experimental investigations — including the far-infrared optical investigations carried out by the present author and his colleagues [44] and reported below — established many properties which were in good agreement with the hypothesis of the existence of a condensed electron—hole phase in germanium. Some discrepancies were found in the estimates of the equilibrium concentration fo particles in the liquid phase n_0 and in the estimates of the average concentrations at which the effects associated with the presence of the condensate were observed. For example, according to [39], the appearance of the condensed phase was observed at $\bar{n} \gtrsim 10^{15}$ cm^{-3} and the equilibrium particle concentration in the liquid phase was $n_0 \approx 10^{16}$ cm^{-3}, whereas, according to [34, 40, 44], the liquid phase existed at much lower average concentration $\bar{n} \sim 10^{13}$-10^{14} cm^{-3} and the equilibrium concentration was an order of magnitude higher $n_0 \approx 2 \times 10^{17}$ cm^{-3}.

Similar studies of the collective properties of excitons were also carried out on silicon. A new line in the spectra of the low-temperature photoluminescence of pure silicon was studied in [45-49]. All the main experimental results reported in these papers could be explained satisfactorily, as in the case of germanium, by the model of spherical electron—hole drops. In particular, the experimentally determined profile of the luminescence line of silicon was in good agreement [46] with that calculated for an equilibrium particle concentration in drops $n_0 = 4 \times 10^{18}$ cm^{-3} and a study of the influence of uniaxial compression on the luminescence spectrum of silicon [49] confirmed the results obtained in [43] for germanium. It should be pointed out that, according to [46-48], the relationship between the lifetimes in silicon was opposite to that found in germanium: in the case of silicon, the exciton lifetime was 1 μsec and the carrier lifetime in the condensate was 0.1-0.5 μsec. This observation was not explained. It was probable that the reduction in the lifetime of nonequilibrium carriers in the dense condensed phase in silicon ($n_0 = 4 \times 10^{18}$ cm^{-3}) was due to the impact (Auger) recombination of electrons and holes whose cross section was several orders of magnitude higher for silicon than for germanium.

Some of the experimental observations reported above could be explained satisfactorily on the assumption that, at low temperatures, excitons formed biexciton molecules. For example, the appearance of a new line in the photoluminescence spectra of intrinsic germanium and silicon was interpreted in [45, 47] as the biexciton luminescence. The main argument in support of this explanation was the quadratic dependence of the intensity of the new line on the excitation rate.

The behavior of an exciton system in strong magnetic fields was studied in [43, 50, 51] using the low-temperature photoluminescence spectra of germanium. Between 40 and 70 kG, the new luminescence line exhibited a doublet splitting which was considerably greater than the thermal energy kT. This splitting was attributed to the appearance of electron–hole drops in which the carriers were degenerate and had an average kinetic energy comparable with the splitting. The equilibrium concentration of particles in a drop was assumed to be 2×10^{17} cm^{-3} and the appearance of the splitting was attributed to the recombination of electrons and holes belonging to different Landau subbands.

Fluctuations of the current across a germanium p–n junction were investigated in [52] in a situation where electron–hole pairs were excited in the immediate vicinity of the junction. When the average concentration of the optically generated carriers was $\bar{n} \gtrsim 10^{15}$ cm^{-3}, the p–n junction current exhibited giant fluctuations corresponding to the passage of about 10^7 elementary charges. This was attributed to the destruction, by the p–n junction field, of electron–hole drops which reached the high-field region and contained 10^7 particles.

A similar experiment was reported in [53]. This followed an earlier study of the photoluminescence of germanium at helium temperatures [45] and attributed the appearance of a new line to the biexciton luminescence. The results reported in [53] led the authors to the conclusion that electron–hole drops, containing 10^6-10^7 carrier pairs and having geometrical dimensions of the order of 1 μ, appeared in the germanium at low temperatures (2°K). These results were deduced from a study of the statistical distribution of the current fluctuations in a p–n junctions as a function of the amplitude of these fluctuations.

According to the results reported in [52], giant fluctuations of the p–n junction photocurrent were not observed in germanium when the average concentration of electron–hole pairs was below 10^{15} cm^{-3}. Hence, it was concluded that, at average concentrations $\bar{n} \sim 10^{13}$-10^{15} cm^{-3}, the exciton condensate could not form and the phenomena observed in this concentration range (for example, the recombination line at hν = 709 meV) were due to the formation of biexcitons. Several arguments were put forward in [54] in support of this view.

According to [54], the most convincing argument for the biexciton nature of the 0.709 eV line in germanium was the quadratic dependence of the intensity of this line on the intensity of the 0.713 eV exciton line at sufficiently low excitation rates. At high excitation rates, the intensities of the two lines were proportional. The latter explanation was explained in [54] on the basis of the rate equation for the exciton and biexciton concentrations in the presence of nonequilibrium long-wavelength phonons generated in the biexciton formation process. The low-temperature lifetime of these phonons could exceed considerably the exciton lifetime under interband recombination conditions. As a result of the accumulation of nonequilibrium phonons at high excitation rates, the number of these phonons became large compared with the number of thermal phonons and this shifted considerably the equilibrium between the biexcitons and excitons in favor of the latter. Thus, at high excitation rates, the dependence of the biexciton line intensity on the exciton line intensity became linear.

The biexciton binding energy estimated from the temperature dependence of the intensities of the exciton and biexciton lines was approximately 5 meV. In view of the fact that the energy separation between the exciton and biexciton luminescence lines at least did not exceed this value, it was difficult to visualize the biexciton recombination mechanism which would explain the energy position of the biexciton line and, particularly, the position of its short-wavelength wing.

It was concluded in [54] that the metallic phase described earlier by the same authors in [39] consisted not of excitons but of biexcitons. In their opinion, exciton molecules existed in the range of average exciton concentrations $\bar{n} \sim 10^{13}$-10^{15} cm^{-3} and the condensed phase described in [40, 44] could not form at these average concentrations.

The nature of the new line (709.6 meV) in the luminescence spectrum of germanium observed at average nonequilibrium carrier densities in the range 10^{13}-10^{15} cm^{-3} was determined in [41] by investigating the diffusion of excitons and other entities responsible for the new luminescence line. Since the effective masses of excitons and biexcitons should be similar, the diffusion parameters (diffusion length and coefficient) of excitons and biexcitons should also be similar. On the other hand, electron−hole drops, consisting of very large numbers of particles, should have a relatively large mass and small values of the diffusion length and coefficient. To determine the diffusion length of excitons and entities responsible for the luminescence at $h\nu$ = 709 meV, the authors [41] measured the intensities of both lines (714.2 and 709.6 meV) as a function of the distance between a small spot where the excitation was concentrated and the region where the luminescence was recorded.

According to the results obtained, the exciton diffusion length was a reasonable value of 4×10^{-2} cm, whereas the diffusion length of the entities responsible for the line at $h\nu$ = 709.6 meV did not exceed 5×10^{-3} cm (this value was limited by the resolution of the system). Moreover, a moving light-spot method was used to determine independently the lifetimes and diffusion coefficient of excitons and new entities. The lifetimes agreed with those found by other methods: they were 8 and 20 μsec, respectively. The diffusion coefficient of excitons was found to be 1500 cm^2/sec, whereas the diffusion coefficient of the new entities was at least 10 times smaller and could not be measured exactly because of insufficient resolution. An estimate of the diffusion coefficient of the new entities was obtained from the measured lifetime (20 μsec) and the diffusion length ($\leq 5 \times 10^{-3}$ cm) and it was found to have a value not exceeding 1 cm^2/sec, which was at least three orders of magnitude smaller than the diffusion coefficient of excitons. It was concluded that such considerable differences between the diffusion properties of excitons and biexcitons were unrealistic. The results of these measurements indicated that the 709.6 meV luminescence line of germanium, observed at average carrier densities of 10^{13}-10^{15} cm^{-3}, was due to the recombination of electrons and holes forming condensed phase drops. Using the momentum relaxation time of a drop taken from [42], the authors [41] estimated the mass of an electron−hole drop, the number of particles therein ($\sim 10^6$), and its geometric dimensions (of the order of several microns). These estimates agreed with those obtained by other workers [44, 53].

The most convincing experiment demonstrating the existence of electron−hole drops in germanium was the observation of the scattering of light by these drops [55]. The complex permittivity of the condensed-phase drops, governing the interaction with electromagnetic radiation, should differ from the permittivity of the rest of the crystal because of the higher carrier density in the condensate. A crystal containing electron−hole drops should be optically inhomogeneous and should scatter transmitted light. The scattering of light of $\lambda = 3.39 \mu$ wavelength, corresponding to the hole absorption band of germanium, was investigated [55] using pure germanium samples in which nonequilibrium carriers with an average density of 10^{13}-10^{15} cm^{-3} were excited optically. Under these conditions, the light transmitted by a sample at $T = 2°K$ was scattered. The absolute intensity and angular distribution of the scattered light were measured. The scattered-light signal varied with the modulation frequency of the exciting radiation and disappeared when the excitation was stopped or when the temperature was raised. Hence, it was concluded that the observed scattering of light was due to the existence in germanium, at $T = 2°K$ and for average carrier densities of 10^{13}-10^{15} cm^{-3}, of condensed phase drops which disturbed the optical homogeneity of the germanium crystal. A comparison of the experimental and calculated results yielded the dimensions of the scattering particles, found by two independent methods including the absolute intensity and angular distribution of the scattered light. The dimensions of electron—hole drops found by these two methods were in good agreement with each other and with the results obtained elsewhere [41, 44, 53]: depending on the excitation rate, the drop radius varied within the range 3-8 μ.

Thus, a considerable body of experimental data supports the hypothesis that, in a wide range of average carrier densities exceeding 10^{12} cm^{-3}, a condensed phase of electrons and holes exists in germanium at suitable temperatures. However, some investigators are of the opinion that, at moderate average carrier densities $\bar{n} < 10^{15}$ cm^{-3}, the formation of the condensed phase is impossible and they explain some of the observations by the formation of exciton molecules. This explanation fails to account for such observations as the far-infrared absorption and luminescence, behavior of the 0.709 eV recombination line in external perturbing fields (mechanical and magnetic), scattering of near-infrared light, etc.

CHAPTER II

METHODS USED IN FAR-INFRARED INVESTIGATIONS OF EXCITONS IN SEMICONDUCTORS

§1. Spectroscopic Measurements

An experimental investigation of the energy spectrum and collective properties of excitons was carried out by the present author using mainly the far-infrared absorption and luminescence spectra of excited germanium.

Intrinsic germanium crystals are transparent to far-infrared radiation at helium temperature and their transmission is governed solely by the reflection losses. When nonequilibrium carriers are generated in such crystals, one should observe absorption due to the interaction between electromagnetic radiation and free or bound (into excitons or exciton complexes) electrons and holes. This absorption can be deduced from the relative measurements of the transmission of a sample during optical generation of carriers compared with its absorption in the absence of carrier-generating radiation. The long-wavelength infrared spectra of the relative transmission of excited germanium crystals at helium temperatures were determined by a special method, in which the main parts of the apparatus were a low-temperature light-guide unit for optical measurements during carrier generation, a source of exciting radiation, and a DIKS spectrometer. The same slightly modified method was used in optical investigations of a different type, namely, in a study of the long-wavelength infrared luminescence of germanium crystals discovered under the same experimental conditions.

Spectrometer. The DIKS spectrometer was a vacuum long-wavelength infrared instrument based on the Ebert—Fastie monochromator (Fig. 1) and built at the Laboratory of Semi-

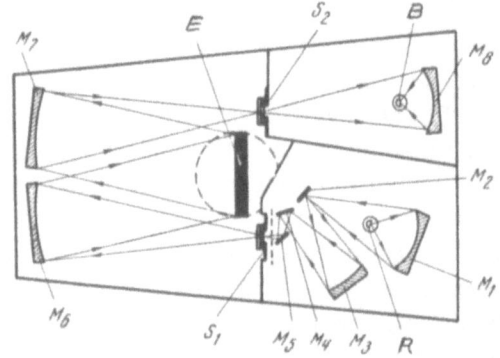

Fig. 1. Optical parts of the DIKS spectrometer: R is a radiation source; E is an echelette grating; M_1, M_3 and M_6-M_8 are spherical mirrors; M_2, M_4, and M_5 are reflection filters; S_1 and S_2 are the entry and exit slits of the monochromator; B is a bolometer radiation detector.

conductor Physics at the Lebedev Physics Institute [56]. Throughout the working range λ = 40-2500 μ the light source was a mercury-in-quartz lamp of the PRK-4 type, operated above its nominal voltage. The dispersive element in the spectrometer was an interchangeable echelette diffraction grating, whose large area ensured a high luminosity of the spectrometer. The absence of a preliminary monochromator made it very necessary to filter the radiation of a given wavelength λ from higher orders ($\lambda/2$, $\lambda/3$, etc.) because, after diffraction by the echelette, these orders traveled in the same direction as the main beam. Various combinations of absorption and reflection filters were used, depending on the spectral region. Paraffin polyethylene, Teflon (usually covered with soot), crystalline and fused quartz, etc. were used as the absorption filters. The reflection filters were small echelettes operating in zero diffraction order, crystals reflecting in the reststrahlen range, and metal wire grids.

The filtration unavoidably reduced the intensity of the radiation reaching the spectrometer and this lowered the sensitivity of the method. This problem was particularly serious at the longest wavelengths because of the absence of efficient filters and because of a strong reduction in the intensity of the mercury-lamp radiation with increasing wavelenth. For this reason, an attempt was made to extend, in the direction of longer wavelenths, the use of metal wire grids whose filtering properties at shorter wavelengths were known to be better than those of absorption and reflection filters. A study of the filtering properties of metal grids with different parameters and structures and a comparison of the results with calculations enabled us to prepare and use successfully high-efficiency reflection grid filters in the wavelenth range up to 2500 μ [57]. Figure 2 shows the spectral characteristics of such filters. It is clear from this figure that the grid filters had a sharp reflection edge, low reflection up to the edge, and nearly 100% reflection beyond it. The use of metal grids in the wavelength range $\lambda \geq 250$ μ, which was very important in the present study, enabled us to increase considerably the sensitivity of the measurements.

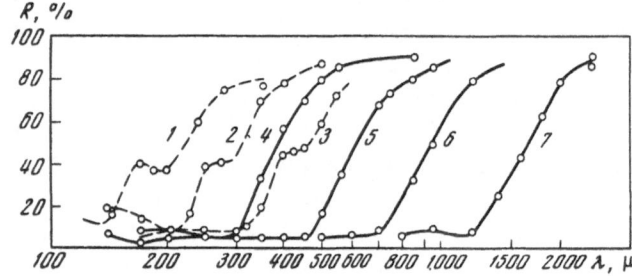

Fig. 2. Reflection spectra of metal grids with different periods (μ): 1) 80; 2) 135; 3, 4) 215; 5) 354; 6) 500; 7) 1000.

The long-wavelength infrared radiation emerging from the spectrometer was directed
to a light-guide cryostat by an evacuated conical guide. The path of the light beam in the
atmosphere and associated radiation losses due to absorption by water vapor were thus practically zero.

§2. Apparatus Used in Low-Temperature Optical

Measurements under Interband Excitation Conditions

Specific difficulties are encountered in the far-infrared spectroscopic measurements
and these are particularly great when the crystal under investigation is subjected to a strong
optical excitation. It is difficult to measure the optical transmission at low temperatures
primarily because of the weak intensity of the probing radiation and relatively large apertures
of the light beams used in the far-infrared range. These circumstances arise from the absence of sufficiently efficient far-infrared radiation sources. Even greater difficulties are
encountered in spectroscopic measurements of the long-wavelength infrared luminescence
emitted by excited germanium because the spectral density of this luminescence does not exceed 10^{-9} W/cm.

Under these conditions, there are considerable advantages in the use of light-guide systems
which make it possible to reduce a light beam to a relatively small transverse cross section
without significant losses and then to measure such a beam directly at the temperature of the
cooling agent. In apparatus of this kind, it is convenient to place a low-temperature infrared
radiation detector inside a cryostat. Detectors of this kind have important advantages, compared with detectors operating at room temperature, of low noise and high threshold sensitivity. The most convenient type of detector, which is suitable for the majority of investigation
in the far infrared when a fast response is not needed, is the low-temperature germanium bolometer. This detector is nonselective in a wide range of infrared wavelengths and its threshold
sensitivity is not inferior to the best low-temperature detectors of other types.

We used a light-guide system specially developed for low-temperature optical ($\lambda = 40$–2500 μ)
measurements on optically excited crystals. The apparatus included a light-guide optical

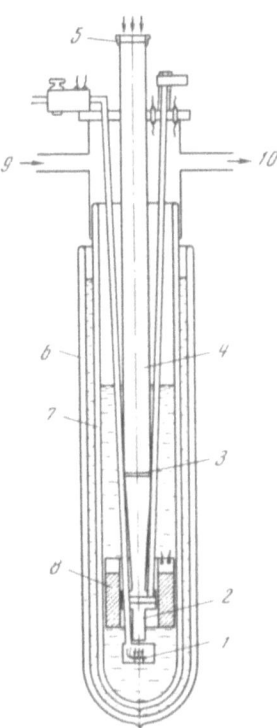

Fig. 3. Construction of a light-guide cryostat: 1)
radiation detector (germanium bolometer); 2-5)
light guide with cooled filters; 6) nitrogen Dewar;
7) helium Dewar; 8) superconducting magnet; 9)
from atmosphere; 10) to pump.

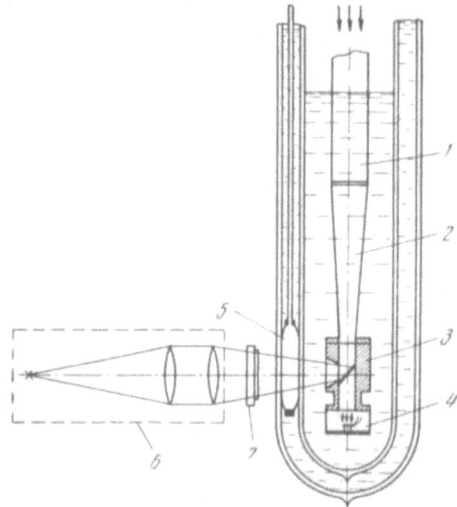

Fig. 4. Experimental arrangement used in far-infrared range during optical generation of carriers: 1) light guide; 2) cone; 3) sample holder; 4) bolometer chamber; 5) Mylar chamber; 6) excitation source; 7) filter.

cryostat, in which a sample was illuminated simultaneously with long-wavelength and exciting radiation and a long-wavelength radiation detector, which was a germanium bolometer [58] placed inside the same cryostat.

The cryostat construction is shown in Fig. 3. The long-wavelength infrared radiation arriving from the spectrometer crossed the entry window of the cryostat and was directed to the bolometer by a light-guide tube whose lower part condensed the beam to a cross section of 9 mm diameter with the aid of a light-guide cone. A unit for mounting and moving a sample (Fig. 4) was attached to the cone; the main element of this unit was an insert with two light-guide channels. In absolute measurements of the "dark" transmission, a sample was placed in one of the channels, whereas, in relative measurements under excitation conditions, it was possible to study two samples employing both channels. The samples inside the insert were placed at an angle of 45° with respect to the light-guide axis and this made it possible to illuminate them simultaneously with the long-wavelength (along the guide axis) and exciting (through a 8-mm-diameter side aperture in the guide) light. The insert could be moved from outside in order to place the samples in the light guide. A miniature device, used to produce a strong local deformation by a steel needle, was mounted on the insert.

The radiation transmitted by the sample passed through the window of an evacuated bolometric chamber and was directed by a short cone to an integrating cavity containing the germanium bolometer. Various absorption filters (black polyethylene, black paper, Teflon, fused quartz, etc.), kept at a low temperature, was placed inside the light guide in front of the bolometer window to remove the background radiation and prevent the exciting radiation from reaching the bolometer. The long-wavelength radiation losses were reduced by silvering and polishing all the internal surfaces in the light guides.

The main parameters of the bolometer at 1.5°K were as follows: resistance 100 kΩ-2 MΩ, sensitive area 6×5 mm, responsivity 2×10^4 V/W, time constant 10^{-3} sec, threshold sensitivity 3×10^{-12} W$\cdot$sec$^{1/2}$. At T > 2°K, the bolometer noise rose somewhat and the sensitivity decreased.

In measurements of the relative transmission of excited samples, the investigated crystal and the bolometer were usually placed in the same cryostat. The influence of fluctuations of the liquid helium temperature on the bolometer, which was due to the strong illumination at temperatures above the λ point of helium, was reduced by protecting the bolometer chamber with a thick Teflon screen. In some experiments, the influence of the exciting radiation on the bolometer was suppressed completely by placing the sample and bolometer in different cryostats joined by a bent light guide. The long-wavelength infrared radiation losses in such a bent guide were quite high and this resulted in a severalfold reduction in sensitivity.

In studies of the long-wavelength infrared luminescence of optically excited germanium, the spectrally unresolved luminescence emerged from a sample inside a light-guide cryostat, passed through an evacuated bent light guide, and reached the DIKS monochromator from which it passed to a second light-guide cryostat with a germanium bolometer.

The bolometer was connected to the input of a low-noise narrow-band amplifier with the usual circuit. The amplifier, whose entry stage represented a cascode circuit with 6S3P vacuum tubes, had an input noise voltage below 3×10^{-8} V, i.e., it did not reduce the sensitivity of the whole system. The amplifier was tuned to the modulation frequency of the long-wavelength infrared luminescence, which was 18 Hz, and the pass band of the amplifier was 1 Hz. In measurements of the transmission spectra, we used the same modulation frequency for the long-wavelength infrared radiation emerging from the spectrometer, whereas in studies of the luminescence emitted by excited germanium we employed a differential method in which the excitation was modulated at this frequency. The amplified signal was passed to a synchronous detector and plotted automatically.

In the case of strong optical excitation of a sample, considerable difficulty was encountered because of the formation of bubbles in liquid nitrogen in the path of the exciting light beam. This difficulty was avoided by displacing liquid nitrogen from the path of the exciting radiation with the aid of a small chamber built from a polyethylene terephthalate sheet. This chamber was placed between the walls of the nitrogen and helium Dewars and filled with gaseous helium kept at a low pressure (Fig. 4).

The apparatus described was very effective, particularly in relative measurements. The most favorable range of temperatures in the case of strong optical excitation was the interval from 2°K down, which corresponded to the distance of the superfluid phase of liquid helium. However, some of the measurements were also carried out at higher temperatures right up to 4.2°K; in these cases, we used bolometers of higher resistance and took several protective measures to suppress the influence of the optical excitation of the sample.

The germanium samples were rectangular wafers with large dimensions ranging from 24×16 to 9×6 mm. This enabled us to fill completely the light-guide cross section. Optical interference was eliminated by beveling the samples at an angle of 1°. The average thickness ranged from several tens of microns to 1 mm. The sample surfaces were ground and etched (immediately before measurements) in a hydrogen peroxide–alkaline mixture in order to reduce the surface recombination velocity.

We studied crystals of intrinsic germanium with a residual impurity concentration $N_i \sim 10^{12}$ cm^{-3} (the residual impurity was aluminum in the case of p-type samples and phosphorus in the case of n-type samples) as well as crystals doped with various shallow impurities (arsenic, phosphorus, antimony) in concentrations ranging from 10^{14} to 10^{16} cm^{-3}.

§3. Sources of Exciting Radiation

Since excitons in a semiconductor crystal have a finite lifetime, a sufficiently high exciton concentration has to be maintained in order to investigate the far-infrared spectra. Obviously, the creation of one exciton requires an energy slightly smaller than the forbidden-band width. The simplest exciton generation method is the optical excitation of electron–hole pairs by the absorption of photons of energy $h\nu \geq E_g$.

At sufficiently low temperatues $kT \ll \mathcal{E}_{ex}$, nonequilibrium electrons and holes, which cool in 10^{-8}–10^{-10} sec because of collisions with phonons, are bound into excitons before (in $\sim 10^{-9}$ sec) they can recombine. Some of the energy evolved as a result of exciton formation is transferred to the crystal lattice and the rest increases the kinetic energy of excitons. The mean free time (relaxation time) of excitons scattered by phonons at low temperatures (T ≲ 4°K) is of the same order as that of free electrons and holes, i.e., 10^{-8}-10^{-10} sec. The exciton lifetime, at least in the case of indirect excitons in germanium, is several orders of magnitude longer: $\tau_{ex} \sim 10^{-5}$-10^{-6} sec at T ~ 4°K. Consequently, during its lifetime, an exciton may collide $\sim 10^3$ times with phonons and thus reach equilibrium with the crystal lattice. A concentration gradient, which appears because of the inhomogeneous absorption of $h\nu > E_g$ photons with depth, causes the excitons to travel from the surface into the crystal to depths of the order of their diffusion length L_{ex} ($L_{ex} \sim 0.1$ cm in germanium [41, 59]).

Thus, in the case of interband optical excitation, we find that, in the bulk of a semiconductor at a depth equal to the diffusion length L_{ex}, we have an exciton system which is in thermodynamic equilibrium with the crystal lattice and in which the average concentration is $\bar{n}$. Under steady-state conditions, this concentration is governed by the generation rate of electron—hole pairs G and the exciton lifetime τ_{ex}.

We shall estimate approximately the rates of carrier generation and, consequently, the power of the exciting radiation source needed to ensure that far-infrared exciton spectra are observed for germanium. If we assume that the binding energy of indirect excitons in germanium is $\mathscr{E}_{ex} = 3.5 \times 10^{-3}$ eV and that their reduced mass is $m_r = 0.08m_0$, we find from the hydrogen-like model that the absorption cross section of excitons in their photoionization range is $\sigma_{ex} \sim 10^{-13}$ cm². Consequently, significant absorption is observed even when the generation rate of carriers in a sample is

$$G = \frac{ad}{\sigma_{ex}\tau_{ex}}S \sim 10^{17}\text{—}10^{18} \text{ sec}^{-1} \tag{8}$$

if $\tau_{ex} \sim 10^{-5}$-10^{-6} sec, the excitation area is S ~ 1 cm², and the optical density of the sample is $ad \sim 0.1$. These generation rates of electron—hole pairs in germanium can be ensured by a radiation source emitting $h\nu \sim 1$ eV quanta and having an output power

$$P_{rad} = G\frac{h\nu}{\eta(1-R)} \sim 10^{-1} \text{ W.} \tag{9}$$

Here, $\eta \approx 1$ is the quantum efficiency of the photoionization of germanium for $h\nu \sim 1$ eV; R = 0.36 is the reflection coefficient of germanium. Very-high-pressure mercury and xenon lamps, characterized by different output powers and used in combination with thick water filters, were used as the excitation radiation sources. We also employed an LG-106 argon laser ($\lambda_{rad} \approx 0.5 \mu$) with a cw output power of up to 1 W as well as various incandescent lamps.

The best results were obtained using a 100-W incandescent lamp of the STs-62 type. The light from this lamp was focused by a system of lenses into a spot whose diameter was 8 mm on the surface of the sample. The long-wavelength part of the lamp radiation with $\lambda > 1.6 \mu$, which did not participate in the generation of electron—hole pairs, was removed by a thick (~15 mm) glass plate and KDP crystal [60]. The radiation intensity of this excitation source was measured with an IMO-1 calorimeter and could be varied continuously from several milliwatts to 0.8 W (Fig. 5). The exciting radiation was attenuated with calibrated metal grids. The dependence of the carrier generation rate on the excitation power was deduced from the room-temperature photoconductivity $\Delta\sigma_{ph}$ of germanium:

$$G = \frac{\Delta\sigma_{ph}}{e\tau_{300° K}(\mu_e + \mu_h)}. \tag{10}$$

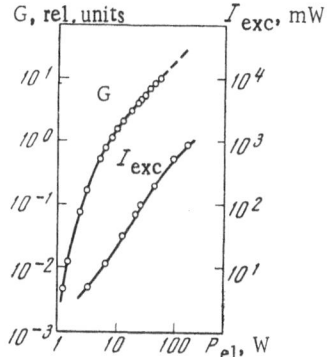

Fig. 5. Dependence of the exciting radiation intensity I_{exc} and of the carrier generation rate G in germanium on the electrical power consumed by the excitation source (lamp).

Here, $\tau_{300°K}$ is the lifetime of electron–hole pairs at T = 300°K; μ_e and μ_h are the electron and hole mobilities at T = 300°K. The results of the measurements are plotted in Fig. 5, where G is the carrier generation rate per unit volume. The low efficiency of the other excitation sources was evidently due to the presence, in their emission spectra, of a large proportion of high-energy photons absorbed in a thin surface layer. In view of surface recombination, the carriers excited in this layer could not contribute significantly to the enhancement of the exciton concentration.

An estimate of the maximum average carrier density generated by the selected excitation source gave a value of the order of 10^{15} cm^{-3}. Assuming that the Bohr radius of an indirect exciton in germanium was a_{ex} = 140 Å [20], we found that at this concentration the quantity $(\bar{n}a_{ex}^3)^{1/3} \approx 0.15$ approached unity, i.e., the average distance between the excitons became of the same order of magnitude as their radius. The potential energy of the interaction between the excitons then became comparable with the binding energy of particles in an exciton. Consequently, even at much lower concentration, the energy of interaction between the excitons could be of the same order of magnitude as or even greater than their kinetic energy. Obviously, under these conditions, the properties of the exciton system should be governed primarily by the interexciton interaction and the collective effects reflecting the nature of this interaction should become manifest.

Thus, the selected excitation source made it possible to study the properties of free noninteracting excitons and the collective properties of excitons in germanium.

§ 4. Thermal Conditions

Our far-infrared spectroscopic measurements were carried out on germanium samples immersed in liquid helium so that the temperature of the sample was 1.5-4.2°K. The attainment and maintenance of these helium temperatures required continuous pumping of the helium vapor through a standard pumping-rate regulator. The helium temperature was monitored in two ways: using the helium vapor pressure, measured with a standard vacuum gauge, and using an Allen–Bradley carbon resistance thermometer located in liquid helium at the same level as the sample.

In the case of measurements carried out under strong optical excitation conditions, the temperature of the sample could rise above that of liquid helium in which it was immersed. According to electrical measurements [61], the thermal contact between the sample and liquid helium was lost and the temperature of the sample increased when the power dissipated by the surface of the sample reached a critical value of about 0.5 W/cm^2. In all our experiments, the specific excitation power never exceeded this value because of the large areas of the samples. Nevertheless, several methods were used to monitor the temperature of a sample during strong optical excitation.

In the measurements at temperatures $T \le 2°K$, i.e., in the range of existence of the superfluid phase of liquid helium, the overheating of a sample could be detected visually because it was accompanied by the appearance of a visible bubble "jacket" around a sample. The construction of the light-guide cryostat was such that the sample surface could be viewed through an upper window in the light-guide tube.

The thermal conditions in a germanium sample immersed in liquid helium and subjected to strong optical excitation could be monitored also by recording the temperature dependence of the ratio of the intensities of the B_1 and B_3 absorption lines of shallow donors in germanium [12]. These lines were due to optical transitions from a chemically split (into two levels A_1 and T_1) ground state of an impurity center to the same excited state $(2p \pm 1)$ (the two lines are shown in Fig. 34 in Chap. V). The relative electron populations of the A_1 and T_1 levels changed with rising temperature so that the population of the upper level T_1 increased and, consequently, the intensity ration B_3/B_1 should rise with temperature. These changes should be greatest at temperatures corresponding to the energy gap between the levels A_1 and T_1, which was 0.32 meV for antimony impurities in germanium, i.e., they should be greatest at temperatures of about 4°K. In fact, according to the results reported in [12], the ratio B_3/B_1 was 1 at $T \sim 9°K$, whereas, at 1.5°K, our results indicated that this ratio was $B_3/B_1 = 0.6$ (Fig. 6a). Thus, the absorption-line intensity ratio B_3/B_1 could be used to determine the true temperature of a sample.

We recorded the absorption lines B_3 and B_1 of control germanium samples doped with antimony ($N_{Sb} \approx 1 \times 10^{14}$ cm^{-3}) and this was done at different excitation rates. The control samples had the same dimensions as the investigated intrinsic germanium samples. At all the excitation rates produced by our source, the ration B_3/B_1 for the samples immersed in superfluid liquid helium corresponded to the helium temperature, which indicated that no overheating took place under these conditions. Unfortunately, the measurements could not be carried out at liquid helium temperatures above the λ point because of fluctuations in the helium column in the light guide, which occurred during the excitation of a sample and which caused strong interference. Overheating of the sample took place when measurements were carried out in helium vapor and this was indicated by the rise of the ratio B_3/B_1 (Fig. 6b) with rising excitation rate.

In some experiments, we determined the temperature of a sample directly during strong optical excitation. This was done by mounting on the sample surface a miniature carbon thermometer [62] whose thickness did not exceed 20 μ and which was insulated from the liquid helium by Zapon lacquer. The calibration curve of one of such microthermometers is plotted

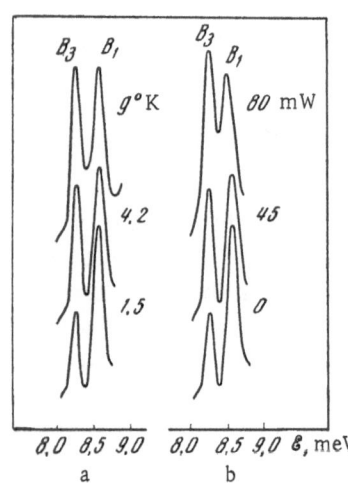

Fig. 6. Impurity absorption lines B_1 and B_3 of antimony in germanium, recorded at different temperatures: a) without excitation; b) at different excitation rates. The sample was kept in liquid helium vapor pumped until 1.5°K was reached.

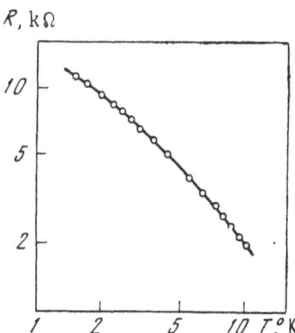

Fig. 7. Calibration curve of a carbon microthermometer.

in Fig. 7. This temperature-monitoring method was used mainly in measurements of the photoionization spectra of free excitons of a sample deliberately overheated relative to liquid helium. In these experiments, the whole apparatus was kept at the optimal temperature ($\sim 1.5°$K) and the sample, insulated thermally from the liquid helium by a thin polyethylene terephthalate film, was overheated by exciting radiation to a temperature of 3-7°K, necessary for the observation of the exciton photoionization spectra.

CHAPTER III

FAR-INFRARED RESONANCE ABSORPTION IN CONDENSED EXCITON PHASE IN GERMANIUM

§1. Absorption Spectra of Intrinsic Germanium

Investigations of the far-infrared absorption in germanium under optical excitation conditions, carried out by the method described in Chap. II, revealed [44] a reduction in the transmission of germanium which occurred at low temperatures (T $\lesssim$ 2°K) over a wide spectral range. In these experiments, the quantity measured directly was the ratio of the transmission of a sample in the case of optical excitation t_G to the "dark" transmission t_0 in the wavelength range 20-1000 μ. The transmission was related to the absorption coefficient by well-known expressions allowing for multiple internal reflections from the surfaces:

$$t_0 = \frac{(1 - R)^2 \exp(-\alpha_0 d)}{1 - R^2 \exp(-2\alpha_0 d)}, \tag{11}$$

$$t_G = \frac{(1 - R)^2 \exp[-(\alpha_0 + \alpha) d]}{1 - R^2 \exp[-2(\alpha_0 + \alpha) d]}, \tag{12}$$

$$t = \frac{t_G}{t_0} = \exp(-\alpha d) \frac{1 - R^2 \exp(-2\alpha_0 d)}{1 - R^2 \exp(-2\alpha_0 d) \exp(-2\alpha d)}. \tag{13}$$

Here, α_0 is the "dark" absorption coefficient; α is the coefficient representing the additional absorption due to the optical excitation; R = 0.36 is the reflection coefficient of germanium; d is the average sample thickness. Interference was eliminated by the use of wedge-shaped samples. The dependence $\alpha d(t, \%)$, calculated in accordance with Eq. (13) for two values of the dark absorption α_0, was of the type shown in Fig. 8. This dependence and the measured values of t_0 and t_G were used to find the absorption (attenuation) αd, which appeared additionally because of the interband optical excitation of the sample.

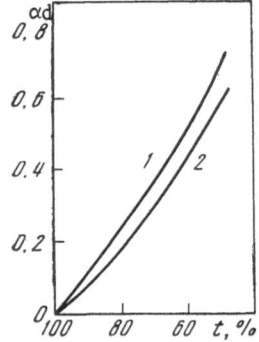

Fig. 8. Dependences of the calculated additional absorption αd on the relative transmission of a germanium sample: 1) $\alpha_0 d = 1$; 2) $\alpha_0 d = 0$.

The results of measurements of the far-infrared absorption of germanium during optical carrier excitation are plotted in Fig. 9. The spectral dependences of the attenuation coefficient αd, measured at 1.5°K, are given for three different carrier generation rates. It is clear from Fig. 9 that the measured spectrum is of resonance nature with an attenuation maximum at the $h\nu = 9$ meV photon energy and a gently sloping short-wavelength wing. The position of the maximum is practically unaffected and the general nature of the spectrum does not change when the excitation rate is varied.

The observed low-temperature resonance obsorption of germanium in the far-infrared under interband optical excitation conditions was interpreted by postulating the appearance of electron—hole drops in germanium [44]. However, before considering the interaction between such drops and electromagnetic radiation, we must discuss briefly other possible mechanisms of the attenuation of long-wavelength infrared radiation in germanium under optical excitation conditions.

The majority of the investigations of far-infrared absorption under interband optical excitation conditions was carried out using very pure germanium with a residual impurity concentration $N_i < 1 \times 10^{13}$ cm^{-3}. In the case of these crystals, we can ignore the contribution of impurity atoms to the attenuation of radiation as a result of optical excitation. However, it should be noted that, in the case of doped and sufficiently strongly compensated samples, such interband excitation may increase the impurity-band absorption because of the capture of nonequilibrium carriers by compensated impurity centers. This effect could play a definite role at certain wavelengths and, therefore, an experimental estimate was obtained of the magnitude of such absorption. The results are quoted in the discussion of the results of measurements carried out on doped samples (§5, Chap. III).

If optically generated electrons and holes are in the free state, the optical excitation may attenuate long-wavelength radiation because of the intraband absorption by nonequilibrium free carriers. According to the classical Drude theory, the free-carrier absorption coefficient

Fig. 9. Spectral dependences of αd, determined at 1.5°K at different excitation rates: 1) 360 mW, G = 14; 2) 230 mW, G = 10; 3) 75 mW, G = 4.

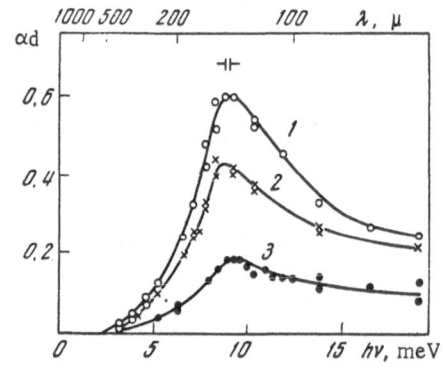

is given by

$$\alpha = \frac{4\pi e^2 n}{\sqrt{\varkappa_0}\, mc} \cdot \frac{\gamma}{\omega^2 + \gamma^2}. \tag{14}$$

Here, n and m are the density and effective mass of the carriers; $\gamma = 1/\tau$ is the effective collision frequency; $\varkappa_0$ is the static permittivity of the investigated crystal; c is the velocity of light in vacuum.

It follows from Eq. (14) that, at low frequencies ($\omega\tau \ll 1$), the absorption coefficient is independent of frequency and proportional to the conductivity, whereas, at frequencies such that $\omega\tau \gg 1$, it varies as $1/\omega^2$, i.e., as λ^2 (λ is the wavelength). The relaxation region corresponding to $\omega\tau \sim 1$ for carriers in germanium at helium temperatures lies in the centimeter range, so that in the investigated part of the spectrum we have $\alpha \propto \lambda^2$. If, using the carrier mobility in germanium at helium temperatures ($\sim 10^6$ cm$^2 \cdot$V$^{-1}\cdot$sec^{-1}), we estimate the effective collision frequency γ and then the absorption cross section σ, we find that, in the region of the maximum in the experimental spectrum, the relevant parameters are $\gamma \approx 2 \times 10^{10}$ sec^{-1} ($\omega\tau = 1$ at $\lambda = 8.5$ cm) and $\sigma \sim 10^{-16}$ cm^2 at $\lambda = 200\ \mu$. Consequently, in the investigated part of the spectrum at the maximum average carrier densities n $\le 10^{15}$ cm^{-3} reached in our investigation, the intraband absorption coefficient of free carriers does not exceed a fraction of 1 cm^{-1}.

There is a different attenuation mechanism due to the presence of free electrons and holes in a crystal. This mechanism involves a homogeneous electron–hole plasma which reflects strongly the radiation of frequencies considerably below the plasma frequency ω_{pl}. In the frequency range $\omega \sim \omega_{pl}$, the reflection passes through a minimum and, in the range $\omega \gg \omega_{pl}$, it is practically constant and governed by the refractive index of the crystal itself. At the average electron–hole concentrations reached in the reported investigation, the frequency range $\omega \sim \omega_{pl}$ lies at wavelengths exceeding 1 mm, i.e., it lies outside the investigated part of the spectrum.

In the experimentally obtained spectra (Fig. 9), the attenuation coefficient falls practically to zero in the long-wavelength range $\lambda = 600$-$800\ \mu$, whereas at the maximum ($\lambda \approx 150\ \mu$), the attenuation is quite strong ($\alpha \sim 10$ cm^{-1}). Thus, neither the nature of the experimentally obtained attenuation spectrum nor its absolute magnitude can be explained by the above interactions between electromagnetic radiation and uniformly distributed free electrons and holes.

The absorption of light by free carriers was determined by measuring the absorption of optically excited germanium at even longer wavelengths up to $\lambda = 2.6$ mm. It was found that, within the limits of the instrumental sensitivity ($\Delta t \sim 0.5\%$), there was no change in the transmission of a sample due to excitation at temperatures T $\le$ 2°K for average electron–hole pair concentrations up to 1×10^{14} cm^{-3}. However, such changes were observed in the transmission of doped samples with antimony in concentrations $N_{Sb} \sim 10^{15}$ cm^{-3} when the measurements were made in helium vapor. The absorption coefficient which appeared due to the optical excitation rose approximately proportionally to λ^2 in the wavelength range 200-800 μ. Measurements of the temperature of a sample during excitation, carried out by the method described in Chap. II, revealed overheating of the sample sufficient for the ionization of a considerable proportion of the shallow donor centers. This accounted for the absorption by free electrons under these conditions.

The statistics of free and bound (into excitons) electrons and holes in germanium-type crystals [63] indicated that at temperatures T $\sim$ 2°K the electrons and holes should be mainly in the bound state. The long-wavelength infrared absorption spectra of free excitons in ger-

manium were investigated for the first time by the present author and his colleagues (see Chap.
V). It was found that the main absorption band was located in the photon energy range 2-5
meV, which was close to the exciton binding energy $\mathscr{E}_{ex} \approx 4.0$ meV. Moreover, this investigation established that the free-exciton absorption of germanium appeared clearly at temperatures of 3-6°K, when the resonance absorption with a maximum at 9 meV was practically
absent. At lower temperatures, these free excitons contributed to the long-wavelength edge
of the resonance absorption and this contribution decreased rapidly when the temperature was
lowered. These results confirmed the hypothesis that the resonance absorption in germanium
at T ≤ 2°K was associated not with the properties of individual excitons but with the collective
properties of the exciton system as a whole.

It was pointed out in Part 2, Chap. I, that the interaction between excitons could not be
regarded as weak even at relatively low average concentrations. One of the results of this
interaction may be the formation of exciton molecules (biexcitons). In spite of the fact that
several theoretical calculations (discussed in Chap. I) have raised doubts about the possibility
of observing biexcitons, at least in crystals with similar effective electron and hole masses,
the resonance absorption observed in the present study could be explained by the biexciton
hypothesis. Additional features of the spectral dependence of the attenuation could also be
explained by this hypothesis but many other experimental observations would be very difficult
to account for on the basis of the biexciton model.

Another important result of the low-temperature interaction in an exciton system with
a high particle concentration may be the formation of a new phase in the form of the electron–
hole condensate. As shown in [44], the resonance absorption found in the far-infrared can
be described satisfactorily on the assumption that the excitons in germanium condense into
electron–hole drops. Further investigations of the observed resonance absorption [64, 65] gave
new experimental evidence which fitted well the drop model but could not be explained easily
by the biexciton hypothesis.

We shall now consider the interaction of electron–hole drops in a crystal with electromagnetic radiation. The perturbation of the electromagnetic field caused by drops is due
to the deviation of the properties of the drops from those of the surrounding medium and due
to their relative dimensions. Each of these factors can be described by a dimensionless parameter. All the optical properties of a drop of interest to us are governed completely by its
complex permittivity $\varkappa$. Since the permittivity of the crystal (medium) shows practically no
dispersion in the investigated frequency range, it is convenient to introduce a relative complex
permittivity of a drop $\tilde{\varkappa}(\omega) = \varkappa(\omega)/\varkappa_0$, where $\varkappa_0$ is the static permittivity of germanium.
It is natural to assume that electron–hole drops are spherical and to describe their relative
size quantity $\rho = 2\pi r/\lambda'$, where r is the average drop radius and $\lambda' = \lambda/\sqrt{\varkappa_0}$ is the electromagnetic wavelength in the crystal. In order to determine the spectral dependence of the attenuation coefficient of a crystal containing electron–hole drops, we must know the frequency
dependence of the permittivity $\varkappa(\omega)$ of the condensed phase and allow for the change in the
relative size of the drop $\rho(\omega)$ with increasing frequency.

Since the theory of the electron–hole liquid in semiconductors is not yet available, the
dependence $\varkappa(\omega)$ can be determined using theoretical representations developed in the electron quantum liquid in metals [66]. The average energy of the Coulomb interaction of electrons in such a liquid is of the order of their kinetic energy. Therefore, in this situation, we
can naturally expect the correlation of electrons to influence some properties of the system.
However, many properties of metals are described theoretically by assuming that electrons
are independent particles, behaving as a gas. Experiments show that, in this case, the presence
of correlations between electrons results in a relatively slight discrepancy between theory and
experiment and this discrepancy is within the order of the quantity being determined. If the
optical properties of an electron–hole liquid are described by the model of a free electron–hole

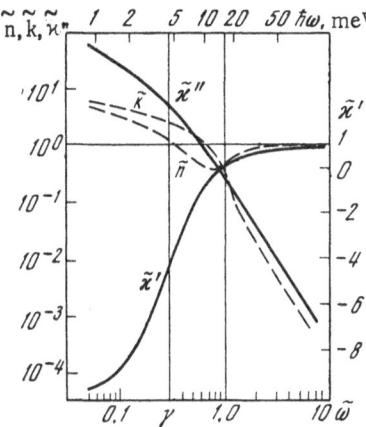

Fig. 10. Dispersion curves of the complex permittivity and refractive index, calculated using Eqs.(15) and (16) for $\gamma = 0.3\omega_{pl}$ = const.

gas, we find that the standard Drude theory yields

$$\varkappa(\omega) = \varkappa_0 - \frac{4\pi e^2 n_0}{m_r \omega (\omega + i\gamma)} = \varkappa_0 - \varkappa_0 \frac{\omega_{pl}^2}{\omega^2 + i\gamma\omega},$$

$$\tilde{\varkappa}(\omega) = \tilde{\varkappa}'(\omega) + i\tilde{\varkappa}''(\omega) = \left(1 - \frac{\omega_{pl}^2}{\omega^2 + \gamma^2}\right) + i\frac{\gamma}{\omega}\frac{\omega_{pl}^2}{\omega^2 + \gamma^2}. \qquad (15)$$

Here, n_0 is the concentration of particles in the liquid; $m_r = m_e m_h/(m_e + m_h)$ is the reduced effective mass; m_e and m_h are the effective electron and hole masses; $\omega_{pl} = (4\pi e^2 n_0/m_r \varkappa_0)^{1/2}$ is the plasma frequency; $\gamma = 1/\tau$ is the effective collision frequency which generally depends on the electromagnetic wave frequency ω.

The dependences of the real $\tilde{\varkappa}'$ and imaginary $\tilde{\varkappa}''$ parts of the permittivity on the relative frequency $\tilde{\omega} = \omega/\omega_{pl}$, calculated in accordance with Eq.(15) for $\gamma = 0.3\omega_{pl}$ = const, are plotted in Fig. 10. This figure gives also the frequency dependences of the refractive index $\tilde{n}(\tilde{\omega})$ and $\tilde{k}(\omega)$ of an electron–hole liquid:*

$$\tilde{m}^2(\omega) = \tilde{\varkappa}(\omega) = [\tilde{n}(\omega) + i\tilde{k}(\omega)]^2,$$

$$\tilde{n} = \text{Re}(\tilde{\varkappa}^{1/2}), \quad \tilde{k} = \text{Im}(\tilde{\varkappa}^{1/2}), \qquad (16)$$

where $\tilde{m}$ is the relative complex refractive index.

It is clear from Fig. 10 that, in the range $\omega \gg \omega_{pl}$, an electron–hole liquid is relatively transparent ($\tilde{k} \ll 1$) and refracts weakly ($|\tilde{n} - 1| \ll 1$), whereas, in the range $\omega \ll \omega_{pl}$, the values of $\tilde{n}$ and $\tilde{k}$ are fairly high, i.e., in this range, the incident radiation is strongly reflected and absorbed by such a liquid.

According to the available data, the average radius of electron–hole drops depends on the excitation range and may amount to one or several microns when the average concentration is $\bar{n} \sim 10^{14}$-10^{15} cm^{-3} [53, 55, 65]. The problem of attenuation of electromagnetic radiation by a medium containing spherical particles of arbitrary size was tackled in the general case by Mie [67], who solved formally the Maxwell equations subject to the appropriate boundary conditions. Attenuation of the transmitted radiation by a single particle was related by Mie to the absorption of energy inside the particle and to the scattering of the radiation out of the original light beam. The formulas for the attenuation coefficient obtained in the general and

_____

*Absolute values of the photon energy in Fig. 10 are given for the experimentally obtained parameters $\omega_{pl} = 2.3 \times 10^{13}$ sec^{-1} and $\gamma = 7 \times 10^{12}$ sec^{-1}.

exact theory were found to be very inconvenient in practical calculations. In our case, the frequency dependence of the attenuation coefficient was found using approximate relationships, valid in certain limiting cases.

The experimental attenuation spectra (Fig. 9) were clearly of the resonance type. This could be due to the specific frequency dependence of the complex permittivity of a particle or due to a definite relationship between the size of a particle and the wavelength of the radiation in the medium. In the former case, the position of the attenuation maximum and the general form of the spectrum are governed principally by the dependence $\varkappa(\omega)$ and do not depend strongly on the particle size, particularly when the condition $\rho \ll 1$ is satisfied and the values of $\widetilde{k}$ are not too small (compared with 1). The attenuation maxima due to the size effect occur for certain values of ρ ($\rho \gtrsim 1$) and are essentially due to the resonance scattering of radiation. Even a slight absorption in the particles themselves reduces oscillations of the attenuation cross section which occur near a certain limiting (for $\rho \gg 1$) value equal to $2\pi r^2$ (πr^2 is the geometric cross section of the particle) and the value of the attenuation cross section at the strongest first maximum [68] never exceeds $6\pi r^2$. Obviously, in this case, the spectral position of the attenuation maximum should vary strongly with the particle size (or, in our experiments, with the excitation rate) because $\rho = \rho(\omega)$.

The experimental spectra shown in Fig. 9 fit better the former of the two situations considered above: when the excitation rate (and, consequently, the size of the electron–hole drops) is varied, the maximum in the spectrum retains its positions and the general form of the curve does not change significantly. Hence, we may assume that the resonance form of the attenuation spectrum is due to the frequency dependence $\varkappa(\omega)$ (see Fig. 10) and the condition $\rho \ll 1$ is satisfied quite well over a major part of the investigated spectral range. If we assume that $\rho \ll 1$, i.e., that the size of the drops is much less than the radiation wavelength in germanium, we find an expression for the frequency dependence of the attenuation coefficient and this expression can be used to show that the dependence is of the resonance type, which is due to the dependence $\varkappa(\omega)$ of the type represented by Eq.(15).

As soon as the condition $\rho \ll 1$ is satisfied, it follows from the frequency dependences $\widetilde{n}(\omega)$ and $\widetilde{k}(\omega)$ (Fig. 10) that the other condition $|\widetilde{m}|\,\rho \ll 1$ is also satisfied. These two conditions represent the Rayleigh case, in which the attenuation cross section of one drop can be represented by a series of powers of ρ [68, 69]:

$$\sigma = \frac{\lambda'^2}{\pi}\,\mathrm{Im}\left[\rho^3\frac{\widetilde{\varkappa}-1}{\widetilde{\varkappa}+2} + \rho^5\frac{1}{15}\left(\frac{\widetilde{\varkappa}-1}{\widetilde{\varkappa}+2}\right)^2\frac{\widetilde{\varkappa}^2+27\widetilde{\varkappa}+38}{2\widetilde{\varkappa}+3} + \rho^6 i\,\frac{2}{3}\left(\frac{\widetilde{\varkappa}-1}{\widetilde{\varkappa}+2}\right)^2 + \cdots\right]. \tag{17}$$

The above general formula for the attenuation cross section is a function of ρ and $\widetilde{\varkappa}$ and is valid in the case of small particles (the electric field is homogeneous within a particle) to within terms whose order exceeds that of ρ^6. The terms in the series (17) describe the dipole-electric, dipole-magnetic, quadrupole-electric, and other interactions of a drop with the electromagnetic field. Obviously, in the case of absorbing particles, the smallness of ρ results in the predominance of the dipole-electric absorption, which is proportional to ρ^3:

$$\sigma_a = \frac{\lambda'^2}{\pi}\,\mathrm{Im}\left(\rho^3\frac{\widetilde{\varkappa}-1}{\widetilde{\varkappa}+2}\right). \tag{18}$$

As ρ rises, we must also allow for the contribution of other terms of the series, particularly that of the dipole-electric (Rayleigh) scattering:

$$\sigma_s = \frac{\lambda'^2}{\pi}\,\frac{2}{3}\rho^6\left|\frac{\widetilde{\varkappa}-1}{\widetilde{\varkappa}+2}\right|^2. \tag{19}$$

The contribution of the dipole-magnetic and other terms can be ignored in the $|\tilde{m}|\,\rho \ll 1$ case.

Substituting in Eqs. (18) and (19) the expression (15) for $\tilde{\varkappa}(\omega)$, we obtain the frequency dependences of the dipole-electric absorption and dipole-electric scattering cross sections of electron–hole drops:

$$\sigma_a(\omega) = V_d\,\frac{3\sqrt{\varkappa_0}}{c}\,\omega_0^2\,\frac{\gamma\omega^2}{(\omega_0^2-\omega^2)^2+\gamma^2\omega^2}\,, \tag{20}$$

$$\sigma_s(\omega) = V_d^2\,\frac{3\varkappa_0^2}{2\pi c^4}\,\omega_0^4\,\frac{\omega^4}{(\omega_0^2-\omega^2)^2+\gamma^2\omega^2}\,, \tag{21}$$

where $\omega_0 = \omega_{pl}/\sqrt{3}$, $V_d = (4/3)\pi r^3$ is the volume of a drop, and c is the velocity of light in vacuum.

It follows from Eqs. (20) and (21) that both the dipole-electric absorption and the dipole-electric scattering exhibit resonances and that the absorption maximum lies at a frequency ω_0, whereas the scattering maximum is shifted slightly away from ω_0 in the direction of higher frequencies:

$$\omega_a^{max} = \omega_0, \tag{22}$$

$$\omega_s^{max} = \omega_0\sqrt{\frac{1}{1-\frac{1}{2}\left(\frac{\gamma}{\omega_0}\right)^2}} = \omega_0\beta, \tag{23}$$

$$\beta \geqslant 1 \quad \text{for} \quad \gamma \leqslant \omega_0.$$

The attenuation coefficient of a medium containing f drops in 1 cm^3,

$$\alpha = (\sigma_a + \sigma_s)\,f, \tag{24}$$

has a maximum at a frequency $\omega_0 \leq \omega^{max} \leq \omega_0\beta$. Obviously, the presence of an attenuation maximum and its energy position are due to oscillations of a drop of frequency ω_0, which is governed by the plasma oscillations of electrons and holes. The absorption coefficient is proportional to the total amount of matter contained in the drops in the path of a beam $\alpha_a \propto V_d f$ and if this mass is constant, it is independent of the drop size. The scattering coefficient $\alpha_s \propto V_d^2 f = (4\pi/3)(V_d f)r^3$ rises strongly with the drop size.

In the simplest case, if the effective collision frequency γ is independent of the radiation frequency ω, the expression (20) for the dipole-electric absorption describes satisfactorily the attenuation spectrum (Fig. 11) [65], with the exception of the short-wavelength range. Comparison of the experimental data with the results of calculations based on Eq. (20) makes it

Fig. 11. Spectral dependence of αd at 1.5°K for an excitation rate of 360 mW. The points are the experimental values. The curves are the results of calculations based on Eq. (20): 1) results obtained on the assumption that γ = const; 2) results obtained on the assumption that $\gamma(\omega) \propto \omega^2$.

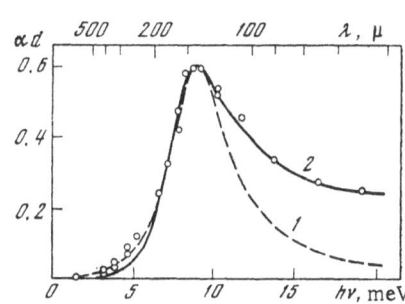

possible to obtain preliminary estimates of some of the properties of electron–hole drops. The most interesting property is the equilibrium density of particles in a drop, which can be estimated from the plasma frequency:

$$n_0 = \frac{m_r \varkappa_0}{4 \pi e^2} \omega_{\text{pl}}^2 = \frac{m_r \varkappa_0}{4 \pi e^2} 3\omega_0^2. \tag{25}$$

The plasma frequency deduced from the experimental data of Fig. 11 is $\omega_{\text{pl}} = \omega_0 \sqrt{3} = 2.36 \times 10^{13}$ sec^{-1}. The equilibrium density of particles in a drop is $n_0 = 2.42 \times 10^{17}$ cm^{-3} if we assume the following parameters for germanium: $m_e = 0.12m_0$, $m_h = 0.30m_0$, and $\varkappa_0 = 16$. Since the values of ω_{pl} and n_0 are deduced from the position of the attenuation maximum, the discrepancy between the calculated and experimental spectra in the short-wavelength range should not affect significantly the validity of the above estimates. The experimental results can also be used to determine the effective collision frequency $\gamma = 6.8 \times 10^{12}$ sec^{-1}, which leads to the damping of plasma oscillations in a drop (this frequency is deduced from the form of the spectrum). The absolute value of the absorption determined experimentally can be used to find directly the total number of electron–hole pairs in drops, which is $N_d = 1.2 \times 10^{13}$ when the excitation power is 360 mW. The average (over the volume of the sample) concentration of the condensed carriers is $\bar{n} = 3.3 \times 10^{14}$ cm^{-3}, i.e., the drops occupy a fraction of the total volume of the sample amounting to $\bar{n}/n_0 = 1.3 \times 10^{-3}$. The constancy of the energy position of the attenuation maximum observed for different excitation rates (Fig. 9) is evidence of the constancy of the particle concentration which is a typical property of the condensed phase. The absolute value of the attenuation rises with the excitation rate because of the increase in the volume occupied by the condensed phase.

We have to know the size of electron–hole drops in order to estimate the contribution of the scattering to the total attenuation of the radiation. In our investigation [44], the contributions of the absorption and scattering to total attenuation were deduced from the best agreement between the experimental results and the calculated values (making the same assumption, $\gamma = $ const, as before) throughout the investigated spectral range up to the photon energy of 20 meV ($\lambda \geq 60 \, \mu$). This agreement was obtained on the assumption that the contributions of the scattering and absorption near the attenuation maximum were approximately equal, which corresponded to the drop radius $r \approx 6 \, \mu$. In this case, the calculated curve described well the measured spectral dependence throughout the investigated range, including the short-wavelength wing. However, it should be pointed out that, for the drop radius $r = 6 \, \mu$, the parameter ρ became equal to unity at $\lambda = 150 \, \mu$, i.e., at the maximum of the spectrum and in the short-wavelength wing we found that $\rho > 1$ so that, in this case, the series expansion (17) was inapplicable.

Some information on the size of electron–hole drops can be obtained from the total amount of the condensate per unit volume. Estimates given above for 360 mW give a fraction $\bar{n}/n_0 = (4/3)f\pi r^3 \approx 10^{-3}$. This is in good agreement with the results of other investigations carried out at similar excitation rates and temperatures, for example, with the results reported in [41, 55] if $n_0 = 2.6 \times 10^{17}$ cm^{-3} and $n \leq 1 \times 10^{14}$ cm^{-3} (the latter value is obtained under excitation conditions similar to those in our study), so that $\bar{n}/n_0 \leq 10^{-3}$. The equilibrium concentration n_0 is obtained in [41] from experimental data of a completely different kind. Our estimate of the average concentration is also supported by a comparison of the value of $\bar{n}$ with the carrier generation rate calculated using the excitation radiation power, which gives the carrier lifetime in a drop $\tau_d \approx 1.6 \times 10^{-5}$ sec, practically identical with the value $\tau_d = 2 \times 10^{-5}$ sec, obtained in [34, 41].

We may thus assume that, at the excitation level in question (360 mW), the fraction of the total volume occupied by the condensate does not exceed 10^{-3}. When the size drop varies and the total amount of the condensate is constant, the drop concentration obeys $f \propto r^{-3}$. The

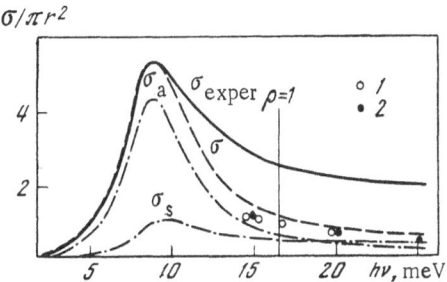

Fig. 12. Calculated spectral dependences of the dipole absorption cross section σ_a, dipole scattering cross section σ_s, and total attenuation cross section σ: 1) results calculated for arbitrary ρ in the "soft" particle approximation; 2) results calculated by the analytic continuation method. The continuous curve represents the experimental data.

attenuation cross section of one drop cannot exceed significantly its geometric cross section and, in any case, we have $\sigma \leq 6\pi r^2$ for particles which are not too small, so that $\rho \sim 1$ [68, 69]. Consequently, for a given amount of condensate and varying radius of the drops, the attenuation coefficient should obey the inequality

$$\alpha \leqslant 6\pi r^2 f = \frac{9}{2}\frac{\bar{n}}{n_0}\frac{1}{r}. \tag{26}$$

It follows from the inequality (26) that the experimentally obtained values of the attenuation coefficient $\alpha_{max} = 15$ cm^{-1} ($\alpha d = 0.6$, d = 0.04 cm, I_{exc} = 360 mW) cannot be explained by the interaction of radiation with particles whose radius exceeds 3 μ.

A possible contribution of the scattering to the total attenuation of the radiation by electron–hole drops can be estimated from Fig. 12, which gives spectral dependences of the dipole absorption σ_a and dipole scattering σ_s cross sections for drops of 3 μ radius, calculated on the assumption that $\gamma = 6.8 \times 10^{12}$ sec^{-1} = const. The same figure gives the dependence of the total attenuation cross section σ, calculated using Eq. (17) with all the terms up to ρ^6, inclusive. All the cross sections are given in units of the geometrical cross section of a drop. We can see that the main contribution to the attenuation is made by the absorption, whereas the scattering does not exceed 15-20% in the region of the maximum and the contribution of other terms in the (17) series is negligible. Practical calculations show [68] that the expansion as a series is highly accurate to $\rho \leq 0.8$, i.e., in the photon energy range $h\nu \leq 12$ meV ($\lambda \geq 100\ \mu$) for drops of radii $r \leq 3\ \mu$. For this reason, the attenuation cross section was calculated for several wavelengths in the short-wavelength region and this was done using different approximate methods valid for particles of arbitrary size, whose complex refractive index differed little from the refractive index of the surrounding medium $|\tilde{m} - 1| \ll 1$ (known as "soft" particles). It is clear from Fig. 10 that, in the case of the postulated dispersion $\tilde{\varkappa}(\omega)$ in the frequency range $\omega > \omega_{pl}$, the conditions $|\tilde{k}| \ll 1$ and $|\tilde{n} - 1| \ll 1$ are really satisfied. Figure 12 gives the results of calculations of the attenuation cross section for several wavelengths in the range $\lambda < 100\ \mu$ obtained by two methods: one is the "soft" particle approximation and the other is the analytic continuation method ($|\tilde{k}| \ll 1$) [68].

Summarizing the results presented in Fig. 12, we may conclude that, if the frequency dependence $\tilde{\varkappa}(\omega)$ is given by Eq. (15), the experimental attenuation spectrum can be explained satisfactorily by the dipole absorption (and, possibly, scattering) of radiation by electron–hole drops of radius $r \leq 3\ \mu$, with the exception of the short-wavelength part of the spectrum. In the case of drops of radius $r \leq 1\ \mu$, it is sufficient to allow only for the dipole absorption throughout the investigated spectral range. Independent measurements of the scattering and absorption would be needed to determine the radius of electron–hole drops and to refine the proportions of the absorbed and scattered radiation.

We shall conclude this section by comparing the cross sections describing the absorption of long-wavelength radiation by a pair of particles present in an electron–hole drop with the absorption cross section of a pair composed of a free electron and free hole. We shall make this comparison in the region of the maximum in the experimental spectrum.

We have already given the absorption cross section of uniformly distributed free electrons and holes present in a crystal in moderate densities ($\bar{n} \leq 10^{15}$ cm^{-3}): $\sigma \sim 10^{-16}$ cm^2 at $\lambda = 200\ \mu$ and helium temperatures.

The condensation produces a strongly inhomogeneous distribution of nonequilibrium carriers because of the formation of dense ($n_0 = 2.4 \times 10^{17}$ cm^{-3}) electron–hole drops occupying a small fraction of the total volume of a sample. The absorption cross section of small (compared with the wavelength) drops has a definite resonance at $\omega_0 = \omega_{pl}/\sqrt{3}$, corresponding to the excitation of plasma oscillations of the drops. In the resonance region, the absorption cross section of a pair of particles in a drop is found experimentally to be $\sigma_0 \approx 5 \times 10^{-14}$ cm^2, i.e., this cross section is more than two orders of magnitude greater than the absorption cross section of a pair of free carriers in the same spectral range.

Thus, the condensation of nonequilibrium carriers into electron–hole drops results in a considerable absorption ($\alpha \sim 10$ cm^{-1}) of the electromagnetic energy in the spectral range corresponding to the frequency of natural oscillations of the drops even when these drops occupy a very small proportion ($\sim 10^{-3}$) of the total volume of a crystal.

§2. Discussion of Parameters of Electron–Hole Drops

(n_0 and γ)

In the preceding section, a comparison of the measured attenuation spectra with the results of calculations in the dipole approximation is used to find some of the parameters representing the properties of electron–hole drops, such as ω_{pl}, n_0, and γ. We shall now consider the meaning of these parameters from the point of view of the electron–hole drop model put forward in [10] and we shall compare them with estimates deduced from the experimental data on the basis of this model in other investigations.

As pointed out in Chap. I, the characteristic length and energy of the condensed electron–hole phase are the radius a_{ex} and binding energy $\mathcal{E}_{ex}$ of individual excitons. When the average distance between particles in a drop is of the order of a_{ex} so that $(na_{ex}^3)^{1/3} \sim 1$, the

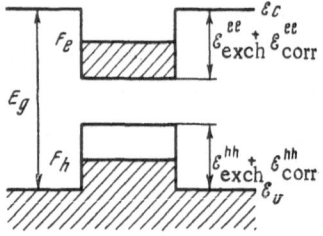

Fig. 13. Schematic representation of the energy bands of a semiconductor after the formation of electron–hole drops.

concentration in the drops in germanium may reach values of the order of 10^{17} cm^{-3}. At these concentrations and helium temperatures, the electron and hole systems are fairly strongly degenerate. The electrons and holes in the condensed phase can be regarded, to some extent, as free because an exciton has, like a univalent atom, a tendency to exchange its electron with neighboring excitons. Consequently, the energy of a system of degenerate electrons and holes decreases because of the exchange and correlation interactions between the carriers. Thus, the forbidden-band width decreases locally at the point where electron–hole drops are located and carriers move at such points in a potential well. The energy scheme of this model is shown in Fig. 13 [34].

The binding energy of a pair of particles in an electron–hole drop may be defined as the difference between the sum of the energies of a free electron and a free hole and the sum of the energies of an electron and hole inside a drop. The total internal energy of a drop can be calculated by summing the average energies of the Coulomb, exchange, and correlation interactions of carriers in a drop with their average kinetic (Fermi) energy. Clearly, the equilibrium state of a drop should correspond to the minimum of the total internal energy. This minimum can be used to find the equilibrium concentration n_0 by varying the concentration of particles in a drop.

The equilibrium concentration n_0 is calculated in [34] by minimizing the energy per pair of particles in a drop. The correlation interaction is allowed for by introducing a correction coefficient (of the order of 3) to the calculated value of n_0. The final value of the equilibrium particle concentration in a drop given in [34] is $n_0 = 2 \times 10^{17}$ cm^{-3}. It is pointed out in [43] that the equilibrium state should correspond to the minimum internal energy of all the particles in a drop. The formula for the internal energy of a drop given in [43] describes correctly the tendency for n_0 to vary in a deformation field but, according to the authors themselves, it is unsuitable for absolute calculations of the equilibrium concentration because of some indeterminacy in the Coulomb energy and because the correlation effects are ignored.

The most reliable calculation of the equilibrium concentration of particles in the condensed phase, based on the minimization of the total internal energy, is given in [70]. It is assumed that, in the case of a homogeneous (on the average) distribution of carriers in a drop, the Coulomb repulsion energy between electrons and holes is compensated by the Coulomb attraction energy between them so that the total internal energy of a drop can be represented in the form

$$E = F_e + F_h + \mathcal{E}_{\text{exch}}^{ee} + \mathcal{E}_{\text{exch}}^{hh} + \mathcal{E}_{\text{corr}}^{ee} + \mathcal{E}_{\text{corr}}^{hh}. \tag{27}$$

Here, $(F_e + F_h)$ is the sum of the average kinetic energies of electrons and holes; $(\mathcal{E}_{\text{exch}}^{ee} + \mathcal{E}_{\text{exch}}^{hh})$ and $(\mathcal{E}_{\text{corr}}^{ee} + \mathcal{E}_{\text{corr}}^{hh})$ are the sums of the average exchange and correlation energies of electrons and holes. The cross terms are ignored in the sums of the exchange and correlation energies. The term $\mathcal{E}_{\text{exch}}^{eh}$ can be ignored because it is proportional to the matrix element of the exchange interaction between an electron and a hole, which is small compared with the matrix element of the exchange interaction between like charges. Clearly, the correlation correction $\mathcal{E}_{\text{corr}}^{eh}$ is of the order of $\mathcal{E}_{\text{corr}}^{ee}$ and $\mathcal{E}_{\text{corr}}^{hh}$, but no calculations of the correlation effects in an electron–hole plasma have yet been published.

The sum of Eq. (27) was calculated in [70] and the equilibrium particle concentration in a drop $n_0 = 3.7 \times 10^{17}$ cm^{-3} was found by equating to zero the pressure at the boundary between the two phases, i.e., between the exciton gas and electron–hole drops in the equilibrium state. In this calculation, the densities of states and the effective masses of electrons and holes in a drop were assumed to be constant.

The equilibrium concentration of particles in a drop was also obtained by comparing the experimental profile of a recombination radiation line of electron—hole drops with the results of calculation based on the model described above [41, 70]. A good agreement between the calculated line profile and the experimental spectrum was obtained for $n_0 = (2.6\text{-}3.0) \times 10^{17}$ cm^{-3}.

An investigation of the recombination radiation emitted by electron—hole drops under uniaxial compression conditions [43, 70] and measurement of the shift of the energy position of the recombination radiation line yielded an estimate of the equilibrium carrier density in a drop at the critical pressure. The value obtained in this way $n_0^{cr} \approx 9.4 \times 10^{16}$ cm^{-3} was in good agreement with the theoretically calculated value $n_0^{cr} = 1.0 \times 10^{17}$ cm^{-3} and demonstrated clearly a tendency for the equilibrium density of the condensed phase to decrease on the application of uniaxial stresses.

Thus, the equilibrium concentration of particles in the condensed phase $n_0 = 2.4 \times 10^{17}$ cm^{-3}, deduced in the present investigation from the position of the far-infrared resonance absorption maximum of electron—hole drops, was in good agreement with the theoretical estimates of this quantity $n_0 = 3.7 \times 10^{17}$ cm^{-3} and with the values of the concentration $n_0 = (2.6\text{-}3) \times 10^{17}$ cm^{-3} calculated from the experimental data on the recombination radiation emitted by such drops. It should be stressed that the far-infrared resonance absorption is due to plasma oscillations of electrons and holes in a drop and the frequency of these oscillations is related directly and unambiguously to the carrier density [see Eq. (25)]. Therefore, our values $n_0 = 2.4 \times 10^{17}$ cm^{-3} can be regarded as reliable.

According to the model described above, all the particles in the dense condensed electron—hole phase are held by internal forces at distances of the order of the exciton radius from one another. The average potential energy of the interaction between particles is then of the order of the average kinetic energy. In this case, the energy of the collective excitation, i.e., the energy of plasma oscillations of electrons and holes, is of the same order of magnitude. In the case of a system of strongly interacting particles (far from the ideal case), we may expect the damping of plasma oscillations to be also of the order of the plasma frequency ($\gamma \sim \omega_{pl}$). These considerations explain qualitatively the considerable width of the observed resonance absorption spectrum and the corresponding value $\gamma \approx 7 \times 10^{12}$ sec^{-1}, deduced from the profile of this spectrum.

We shall now consider the possible mechanisms of the damping of plasma oscillations of electron—hole drops and we shall estimate the corresponding effective collision frequencies. We shall do this using qualitative considerations and some analogies with electrons in a metal because it is practically impossible to provide an exact theoretical description of this case.

We shall consider the interaction of electrons and holes in drops with phonons. At low temperatures (T $\ll \theta$, where θ is the Debye temperature), the momentum exchange is difficult. On the one hand, among the phonons on the Fermi surface there are practically none with momentum of the order of the carrier momentum and the scattering accompanied by phonon absorption is practically absent. On the other hand, the majority of carriers has insufficient energy to emit phonons because the broadening of the Fermi energy of carriers is of the order of kT $\ll$ kθ. Consequently, the mean free time of carriers governed by collisions with phonons τ_{ph}(T) rises very rapidly when the temperature is lowered (at low temperatures this time constant is proportional to T^{-5}) and it is of the order of 10^{-6}-10^{-7} sec for germanium at T = 2-4°K [42]. A similar result is obtained by a classical analysis, which can be applied to the mobility of electrons and holes in a drop.

In investigating the optical properties of the condensed phase in the energy range kT $\ll$ $\hbar\omega \sim$ kθ, we must allow for the quantum properties of the field. In fact, the carriers which have absorbed electromagnetic quanta are far outside the limits of the broadened Fermi distribution. As before, they cannot absorb phonons, but since they now have high energies they can emit phonons in a wide range of the phonon spectrum, limited by the energy of the absorbed quanta (photons). In this sense, the quantum damping mechanism in the range $kT \ll \hbar\omega \sim k\theta$ is similar to the classical damping mechanism at high temperatures. These considerations are confirmed by a quantum-mechanical analysis of the absorption by metals in the visible and near-infrared parts of the spectrum [71], in which case the mean free time under phonon scattering conditions in the range $\hbar\omega \gg k\theta \gg kT$ is described by the expression

$$\tau_{\mathrm{ph}}(T) \approx \frac{5}{2}\tau^{(\mathrm{cl})}(\theta), \tag{28}$$

where $\tau^{(\mathrm{cl})}(T)$ is the usual classical high-temperature mean free time.

In the region of the absorption maximum of electron–hole drops in germanium, the photon energy is $\hbar\omega \approx 9$ meV, which corresponds to $T \sim 100°$K. The mean free time in the case of collisions with phonons at this temperature can be determined from the known carrier mobility in the original germanium at liquid-nitrogen temperature $\mu(77°$K$) \approx 4 \times 10^4$ cm$^2 \cdot$V$^{-1} \cdot$sec^{-1}, which gives $\tau_{\mathrm{ph}}(77°$K$) \approx 2 \times 10^{-12}$ sec. It follows from the discussion given above that we can estimate approximately also the effective collision frequencies of electrons and holes in a drop with phonons at low temperatures and high photon energies $\hbar\omega \gg kT$: $\gamma_{\mathrm{ph}} \approx 5 \times 10^{11}$ sec^{-1}. We may assume that, as the photon energy rises to values of $\hbar\omega \leq k\theta$, γ_{ph} increases in accordance with a law similar to the law describing the fall of the mean free time due to phonon collisions with rising temperature (in the appropriate temperature range), i.e., $\gamma_{\mathrm{ph}}(\omega) \propto \omega^\beta$ if $\tau_{\mathrm{ph}}^{(\mathrm{cl})}(T) \propto T^{-\beta}$, where $\beta \approx 2$.

Thus, the scattering of electrons and holes by phonons in the region of the absorption maximum gives rise to a damping $\gamma_{\mathrm{ph}} \sim 5 \times 10^{11}$ sec^{-1}, which is approximately an order of magnitude smaller than that found experimentally. The impurity scattering should be slight in pure germanium. However, electron–hole collisions may play an important role when the carrier density in a drop is $n_0 \sim 10^{17}$ cm^{-3}. According to the results of electrical measurements [37], electron–hole collisions determine the nature of the conductivity of the condensed phase in germanium. The carrier mobility in the condensed phase ($\mu \sim 10^6$ cm$^2 \cdot$V$^{-1} \cdot$sec^{-1} at $T = 1.7°$K) as well as the temperature and concentration dependences of the mobility [37] are all in good agreement with the mechanism of electron–hole scattering in transition metals [72], where it is found that the frequency of electron–hole collisions is $\gamma_{\mathrm{eh}} \propto (kT/F)^2$. These results can be used to estimate the value of γ_{eh} in the static case: $\gamma_{\mathrm{eh}}^0 = 2.2 \times 10^{10}$ sec^{-1} for $n_0 = 2 \times 10^{17}$ cm^{-3} and $T = 1.7°$K. In the optical investigations, we must allow for the frequency dependence of γ_{eh}. Using the same approach as in the analysis of the interaction between carriers and phonons, we may assume that the dependence $\gamma_{\mathrm{eh}}(\omega)$ in the photon energy range $kT \ll \hbar\omega \ll F$ is close to the law $\gamma_{\mathrm{eh}}^0(T)$, i.e., $\gamma_{\mathrm{eh}}(\omega) \propto \omega^2$.

The quantum theory of the absorption of electromagnetic energy by metals in the electron–electron interaction case gives the following frequency dependence of γ_{ee} in the $kT \ll \hbar\omega \ll F$ range [73]:

$$\gamma_{ee}(\omega,\ T) = \gamma_{ee}^0(T)\left[1 + \left(\frac{\hbar\omega}{2\pi kT}\right)^2\right]. \tag{29}$$

If, continuing the analogy with metals, we use Eq. (29) to estimate the damping in the case of electron–hole collisions in the region of the resonance absorption maximum, we find that an estimate obtained for $T = 1.5°$K gives $\gamma_{\mathrm{eh}} \approx 2 \times 10^{12}$ sec^{-1}, which is close to the experimental value. The expression (29) is derived on the assumption that the energy of electromagnetic quanta (photons) is $\hbar\omega \leq F$. An analysis carried out at higher frequencies $\hbar\omega \geq F$ gives rise to some singularities in the frequency dependence of the absorption coefficient at $\hbar\omega = F$,

$\hbar\omega = 2F$, etc. However, as shown in [73], these singularities are very weak. The most important is located at $\hbar\omega = F$, where the fourth derivative of the absorption coefficient becomes infinite.

The above analysis shows that the effective collision frequency, in the case of interaction between carriers and phonons and between electrons and holes, may increase with rising frequency ω. The frequency dependence $\gamma(\omega)$ explains the short-wavelength asymmetry of the resonance absorption spectrum. Figure 11 shows the resonance absorption spectrum calculated on the assumption that γ is a quadratic function of the frequency. It is clear from this figure that the theoretically calculated curve is in very good agreement with the experimental results. It is likely that the absorption of short-wavelength electromagnetic energy by electron–hole drops is due to transitions between the heavy- and light-hole valence subbands. A consistent allowance for the hole transfer processes should increase the absorption coefficient both directly and indirectly, the latter being due to the contribution to the effective collision frequency γ. However, the mechanism of transitions occurring in the short-wavelength wing of the resonance absorption spectrum of electron–hole drops in germanium cannot be determined without additional experiments because some contribution may come from absorption of different origin.

The above discussion is very qualitative and based on general considerations as well as some theoretical ideas developed for metals. This is unavoidable because we are dealing with systems of particles in which the Coulomb interaction occurs at concentrations $na_{ex}^3 \sim 1$, which cannot be analyzed rigorously.

§3. Temperature Dependence of Resonance

Absorption

One of the most characteristic features of the far-infrared resonance absorption spectrum of germanium under optical excitation conditions is the strong temperature dependence of the absorption. It is reported in [44, 64, 65] that, for a given excitation rate, the resonance absorption is observed only below a certain maximum (threshold) temperature. In a very narrow range of temperatures below this threshold, the absorption rises rapidly when the temperature is lowered and then reaches saturation. The temperature threshold of resonance absorption depends strongly on the excitation rate and shifts toward higher temperatures when this rate is increased. These features of the temperature dependence of the resonance absorption are illustrated in Fig. 14, which gives the experimentally obtained dependences of the product αd at the absorption maximum on the reciprocal temperature at several excitation rates. The curves plotted in Fig. 14 were obtained for intrinsic germanium with a residual impurity concentration $N_i < 10^{13}$ cm^{-3}.

Unfortunately, the range in which such measurements could be carried out was limited to temperatures below $\sim 2.1°$K. This was due to experimental difficulties arising because of the phase transition of liquid helium from the superfluid to the normal state. When the trans-

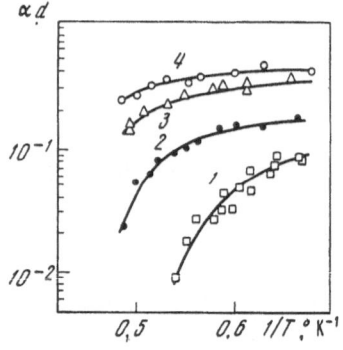

Fig. 14. Temperature dependences of αd, determined at the maximum of the absorption spectrum ($\lambda = 136\ \mu$) at different excitation rates: 1) 40 mW, G = 2.6; 2) 65 mW, G = 3.7; 3) 150 mW, G = 7; 4) 230 mW, G = 10.

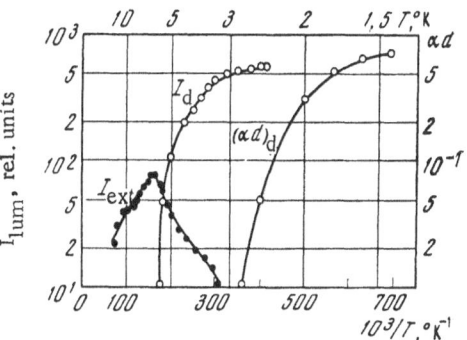

Fig. 15. Temperature dependences of the resonance absorption of frops $(\alpha d)_d$, intensity of the drop luminescence line I_d (709.6 meV), and intensity of the free-exciton line I_{ex} [43].

mission was measured at temperatures above the λ point of liquid helium, the noise due to fluctuations in the liquid helium in the light-guide tube increased strongly. The noise background was so strong that it was often difficult to identify the resonance absorption signal which varied rapidly with rising temperature. It was somewhat easier to identify this signal in the case of doped germanium because, in that case, the resonance absorption was 1.5-2 times higher than in the case of pure germanium excited at the same rate (see §5 in the present chapter).

The existence of a threshold temperature of the resonance absorption, which depended on the excitation rate, i.e., on the carrier or exciton density, was clear evidence of the condensation of excitons in germanium into electron–hole drops. The far-infrared resonance absorption and the new recombination radiation of germanium at $h\nu$ = 709.6 meV, interpreted on the basis of the general prediction of the condensation of excitons into electron–hole drops, had very similar dependences, particularly those on temperature. This similarity is illustrated in Fig. 15, which shows the temperature dependence of the product αd measured for lightly doped germanium (N_{Sb} = 4 × 10^{14} cm^{-3}) as well as the temperature dependence of the intensity of the 709.6 meV recombination radiation line of germanium [43]. The recombination radiation threshold was shifted toward higher temperatures because the crystal was excited by focused He–Ne laser radiation which ensured high average concentrations.

According to theoretical estimates [10], confirmed by experimental investigations reported in [43, 53, 70], the critical (evaporation) temperature of the condensed electron–hole phase in germanium is $T_{cr} \approx 8-10°K$. At temperatures much lower than the critical value, we can expect the liquid phase to appear when the average concentration reaches $n_T \ll n_0$, which depends on temperature and on the difference between the binding energy of a pair of particles in the gaseous and liquid phases. In the gaseous phase, this binding energy is governed by the binding energy of an exciton in the ground state, whereas, in the liquid phase, the binding energy is the total chemical potential of electron–hole drops $F = (F_e + F_h) < 0$, where $F_e < 0$ and $F_h < 0$ are the chemical potentials of electrons and holes measured from the lower edges of the relevant bands. At a given temperature T, the exciton density at which electron–hole drops appear is given by the thermodynamic relationship

$$n_T = \xi \left(\frac{MkT}{2\pi\hbar^2} \right)^{3/2} \exp \frac{F + \mathcal{E}\,ex}{kT},$$ (30)

where ξ and M are, respectively, the order of degeneracy of the ground state and the density-of-states effective mass of an exciton. The expression (30) represents the relationship between the threshold temperature and the threshold excitation rate of the process of exciton condensation into electron–hole drops.

In the range of temperatures well below the critical value, the equilibrium density of the liquid phase and the chemical potentials (Fermi energies) of electrons and holes in this phase are practically independent of temperature. Under these conditions, the quantity F +

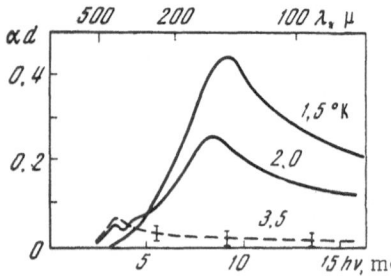

Fig. 16. Spectral dependences of αd at various temperatures. I_{exc} = 230 mW.

$\mathcal{E}_{ex}$ may be found from the experimentally obtained temperature dependence of the threshold excitation rate, in accordance with Eq.(30). However, the difficulties encountered in measuring the resonance absorption near the threshold of the appearance of liquid drops prevent reliable determination of this dependence for two reasons: 1) the temperature interval in the measurement of small values of αd is limited by the λ point; 2) even when the transmission is measured to within 0.5%, it is not possible to determine the moment of appearance of condensed-phase drops in a crystal because of the background absorption in germanium and due to the excitation.

In this situation, there are definite advantages in measuring the intensity of the recombination radiation emitted by carriers in electron–hole drops. The temperature dependences of the excitation threshold of the drop recombination radiation (luminescence) line yielded the values of $F + \mathcal{E}_{ex}$ equal to -2.7 meV [34] and -1.0 meV [70]. The same quantity could be determined from the shift of the recombination radiation (luminescence) lines of free excitons and of drops, as was done in [24, 43, 70]: in this way, it was found that $F + \mathcal{E}_{ex}$ = -2.7 meV. It should be pointed out that the latter value could be somewhat overestimated in view of the fact that the recombination of carriers in a drop resulted in the transfer of some energy to the drop itself and this raised it to an excited state.

We shall use Eq. (30) to estimate the threshold exciton concentration n_T in the temperature range 1.5–4.2°K, where the resonance absorption measurements were carried out. If we take the quantity $F + \mathcal{E}_{ex}$ to have some average values such as -1.5 meV, we find that $n_T(1.5°K) \sim 10^{11}$ cm^{-3} and $n_T(4.2°K) \sim 2 \times 10^{14}$ cm^{-3}. These estimates are not in conflict with the average concentrations corresponding to the excitation source power and values of n deduced from the experimental data on the resonance absorption product αd (§1, Chap. III).

The nature of the absorption spectrum did not vary greatly with temperature below the threshold. Figure 16 shows the absorption spectra obtained at temperatures of 1.5, 2.0, and 3.5°K for 230 mW exciting radiation. The spectral curve obtained at 2.0°K shows some singularities in the photon energy range $h\nu \sim 3$–4 meV. These absorption singularities were observed only for germanium samples with long exciton lifetimes (of the order of several microseconds at helium temperatures) and their magnitudes depended strongly on the surface state of the sample. When the temperature was increased to 3.5°K, the resonance absorption practically disappeared, which indicated that the temperature rose above the threshold value for this particular excitation rate. The singularities in the range $h\nu \approx 4$ meV increased slightly, assuming the form of a wide but weak band (Fig. 16). We shall show in Chap. V that the absorption in this region may be due to free excitons in the gaseous phase.

§4. Dependence of Resonance Absorption on

Excitation Rate

The dependence of the effects observed on the excitation rate was investigated in practically all the studies of the properties of high-concentration exciton systems [34, 35, 45, 51, 54]. This was due to the fact that the kinetics of formation of exciton complexes could give information on their internal structure. For example, if a new photoluminescence line of germanium

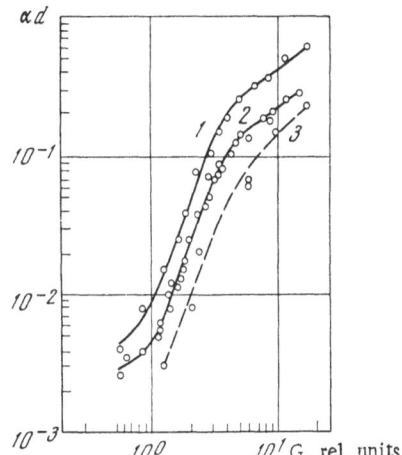

Fig. 17. Dependences of αd on the excitation rate, measured in different parts of the spectrum: 1) λ = 136 μ at 1.5°K; 2) λ = 190 μ at 1.5°K; 3) λ = 136 μ at 1.9°K.

or silicon is due to the annihilation of biexcitons, it follows from the analysis of the kinetics of this process given in [35, 45, 54] that the intensity of such a line at low excitation levels should be proportional to the square of the carrier generation rate (as long as the number of biexcitons remains small compared with the total number of excitons). A stronger (cubic) initial dependence on the carrier generation rate should be observed in the case of recombination of electrons and holes in condensed drops (this applies to the initial stage when the number or carriers forming such drops is small compared with the total number of carriers) [10, 34].

The nature of the resonance absorption was determined by investigating the nature of the dependence of the product αd on the excitation rate at various temperatures using intrinsic and doped germanium samples. This dependence was determined near the maximum of the absorption spectrum and in its long-wavelength wing, where the influence of residual or deliberately introduced impurities was negligible. Typical dependences of the values of αd on the carrier generation rate G are shown in Fig. 17 for intrinsic germanium samples with residual impurity concentrations $N_i < 10^{13}$ cm^{-3}. We can see that at the maximum and in the long-wavelength wing the product αd rises rapidly (αd $\propto G^{2.7}$) at low carrier generation rates, but changes to a nearly linear law (αd $\propto G^{0.7}$) at higher excitation rates. Figure 17 shows the dependences of the maximum absorption on the rate of excitation at two different temperatures. We can see that when the temperature is increased the whole curve shifts toward higher excitation rates, i.e., the excitation threshold depends strongly on temperature.

The relationship between the threshold values of the temperature and particle concentration in the process of exciton condensation into liquid drops has been discussed in connection with the temperature dependence of the resonance absorption on the basis of the thermodynamic relationship (30). However, the distribution of particles between the gaseous phase and liquid drops is not governed simply by thermodynamic relationships because excitons have a finite lifetime and continuous excitation is needed to maintain a given exciton concentration. The number of particles in the liquid phase can be estimated from transport (rate) equations discussed in [10, 34]:

$$\frac{\partial n_{ex}}{\partial t} = G - \frac{n_{ex}}{\tau_{ex}} - \pi r^2 v_{ex}(n_{ex} - n_T) f, \tag{31}$$

$$\frac{\partial}{\partial t}\left(\frac{4}{3}\pi r^3 n_0\right) = \pi r^2 v_{ex}(n_{ex} - n_T) - \frac{4}{3}\pi r^3 \frac{n_0}{\tau_d}. \tag{32}$$

Here, n_{ex}, v_{ex}, and τ_{ex} are the concentration, thermal velocity, and lifetime of excitons in the gaseous phase; n_0 and τ_d are the equilibrium concentration and lifetime of carriers in liquid drops; f and r are the average number of drops per unit volume and their average radius; G is the exciton generation rate (per unit volume).

The average number of drops per unit volume f is governed by the actual drop nucleation mechanism, which is likely to be associated with the presence of condensation centers in real crystals and these centers may be various inhomogeneities, local stresses, pile-ups of defects or dislocations, etc. The number f occurs in Eqs. (31) and (32) as a parameter.

Under steady-state conditions, when $dn_{ex}/dt = dr/dt = 0$, a reduction in the number of electron—hole pairs in a drop as a result of pair recombination is balanced out by the capture of free excitons by the drop. It then follows from Eq. (32) that

$$r = 3 \frac{n_{ex} - n_T}{n_0} v_{ex} \tau_d,$$

(33)

i.e., at concentrations considerably above the threshold value, the average drop radius is proportional to n_{ex}. The concentration n_{ex} is described by a cubic equation which can be obtained from Eqs. (31) and (32) and which has a solution corresponding to positive values of the drop radius only if $G \geq G_T = n_T/\tau_{ex}$. Clearly, G_T is the threshold value of the generation rate at which the condensed phase may appear. At low exciton generation rates, when the recombination of carriers in the exciton gas is the predominant process, it follows from Eqs. (31)–(33) that n_{ex} and r rise proportionally to the generation rate and the volume of the liquid phase increases proportionally to $(G - G_T)^3 \approx G^3$ and f. If the generation rate is sufficiently high, so that carrier recombination occurs mainly in the drops, the total number of condensed carriers is proportional to the generation rate but is independent of f, i.e., the conditions are close to thermodynamic equilibrium.

It is shown in §1 in the present chapter that, in the case of the dipole mechanism of the absorption of electron-magnetic energy by electron—hole drops, the absorption coefficient is proportional to the total number of condensed particles in the path of the beam, i.e., it is proportional to the volume of the liquid phase. Thus, at low excitation levels, the dipole absorption should rise as the cube of the generation rate but the dependence should change to linear at high generation rates. The experimental results obtained for intrinsic germanium indicate that the dependences are close to $\alpha d \propto G^{2 \cdot 7}$ and $\alpha d \propto G^{0 \cdot 7}$. These experimental observations can be regarded as a confirmation of the proposed dipole model of the absorption by electron—hole drops in the plasma oscillation region. A slight deviation of the exponent in the power law from the theoretical predictions may be due to the presence of a constant absorption background due to excitation of the investigated sample.

The nature of the experimental dependence of the attenuation on the generation rate can be used to estimate the contribution of the dipole scattering to the general attenuation of the long-wavelength infrared radiation. We have already mentioned that the dipole scattering should be proportional to the square of the drop volume, i.e., it should be proportional to G^6 at low excitation levels. In fact, the initial region exhibits a weaker dependence $\alpha d \propto G^3$. This means that under the actual experimental conditions employed, the dipole scattering and other forms of attenuation which vary with drop radius as r^β, where $\beta \geq 5$, are quite weak.

The presence of a definite inflection in the experimental dependence of the resonance absorption on the generation rate (Fig. 17) allows us to estimate approximately the average radius of electron—hole drops from the approximate equality (at the inflection point) of the rates of recombination of excitons and of carriers in the drops [65]:

$$r_{inf} = \frac{3}{4} \frac{\tau_\partial v_\partial}{n_0} \bar{n}_{inf}.$$

(34)

The value of n_{inf} corresponding to the inflection point can be found from the measured value of the absorption in the region of the inflection of the dependence $\alpha d\,(G)$. If we use the values $\tau_{ex} \leq 10^{-6}$ sec [34, 45], $n_0 = 2.4 \times 10^{17}$ cm^{-3}, and $v_{ex} \approx 10^6$ cm/sec (at T = 1.5°K), we find that the average drop size is r $\approx 3 \times 10^{-4}$ cm. This estimate is in agreement with the results of other experimental investigations [53, 55] and not in conflict with our assumption that, in the investigated spectral range, the drop size is less than the radiation wavelength. This value of the radius naturally represents only an order-of-magnitude estimate.

Thus, the initial cubic dependence of the resonance absorption on the excitation rate, which becomes linear at higher excitation rates, can be regarded as one of the arguments in support of the proposed dipole model of the absorption of electromagnetic energy by electron–hole drops in the plasma oscillation region. Similar dependences on the excitation rate were reported in [34, 51, 70] also for the intensity of the new luminescence line (709 meV) observed in the low-temperature photoluminescence spectrum of germanium and attributed to the recombination of carriers in condensed electron–hole drops. If the biexciton model is adopted, there is no real justification for assuming the dependence on the excitation rate to be stronger than quadratic and this applies to the long-wavelength infrared absorption and luminescence line intensity.

§ 5. Resonance Absorption in Doped Germanium

The resonance absorption in germanium excited continuously at low temperatures was also investigated using samples specially doped with shallow impurities. The purpose of this investigation was to determine the direct contribution of dopant (or residual) impurities to the long-wavelength infrared absorption in a sample subjected to optical excitation and to determine the possible role of impurities in the formation of condensed phase drops. With these aims in mind, we subjected doped germanium to all the measurements carried out earlier on intrinsic germanium.

Investigation of the absorption spectra of doped germanium samples at temperatures T ≤ 2°K established that the resonance absorption with a maximum at hν = 9.0 meV was observed without significant changes even when the impurity concentration did not exceed $N_i \leq 10^{15}$ cm^{-3}. Similar absorption spectra were obtained for samples doped with different shallow impurities. Figure 18 compares the resonance absorption spectra of intrinsic germanium with $N_i \approx 1 \times 10^{12}$ cm^{-3} and of germanium doped with antimony in an amount $N_{Sb} = 4 \times 10^{14}$ cm^{-3} (the compensating acceptor concentration was $N_A = 2 \times 10^{13}$ cm^{-3}); these spectra were determined at T = 1.5°K and the excitation radiation power was I_{exc} = 230 mW. It is clear from Fig. 18 that the profiles of the spectra and the energy positions of the maxima agreed quite well although the absolute value of the absorption was somewhat higher, for a given excitation rate, in the case of the doped sample. Similar results were also obtained for samples doped with P and As.

This absence of the dependence of the absorption spectra on the nature of the impurity atoms and their concentration up to $N_i \leq 10^{15}$ cm^{-3} indicated that the resonance absorption

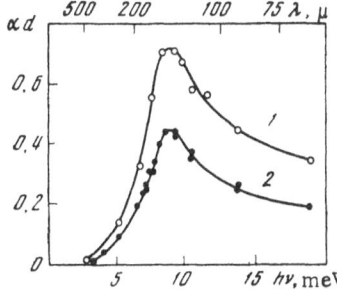

Fig. 18. Spectral dependences of αd, determined at 1.5°K for I_{exc} = 230 mW: 1) for doped germanium with $N_{Sb} = 4 \times 10^{14}$ cm^{-3}; 2) for pure germanium with $N_i \approx 1 \times 10^{12}$ cm^{-3}.

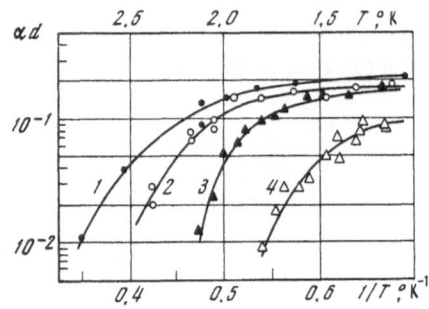

Fig. 19. Temperature dependences of αd, measured at $\lambda = 145\ \mu$: 1, 2) for doped germanium with $N_{Sb} = 3.6 \times 10^{15}$ cm^{-3}; 3, 4) for pure germanium with $N_i \approx 1 \times 10^{12}$ cm^{-3}. Excitation rate: 1, 3) 65 mW ($G = 3.7$); 2, 4) 40 mW ($G = 2.6$).

was not associated with the presence of residual impurities but was a manifestation of the collective properties of the exciton system in germanium.

Investigations of the temperature dependence of the resonance absorption of doped germanium showed that, as in the case of intrinsic germanium, this absorption was observed only below a certain threshold temperature. In this case, again, the threshold temperature rose with excitation rate and, when this rate was kept constant, the threshold shifted in the direction of higher temperatures when the impurity concentration was increased. This is illustrated in Fig. 19, which shows the temperature dependences of the resonance absorption in intrinsic and antimony-doped ($N_{Sb} = 3.6 \times 10^{15}$ cm^{-3}) germanium samples, obtained at two different rates of generation near the absorption maximum. Similar measurements were carried out also in the long-wavelength wing of the spectrum, outside the shallow-impurity absorption band, and they gave similar results.

Investigations of the spectral and temperature dependences of the resonance absorption in doped germanium indicated that the condensed electron—hole phase appeared also in doped samples and the presence of impurities in concentrations not exceeding $N_i \le 10^{15}$ cm^{-3} did not alter significantly the properties of the electron—hole drops. On the other hand, the presence of impurities (in this range of concentrations) increased the threshold temperature of the existence of the condensed phase drops, and, at a fixed temperature and excitation rate, it increased the absorption of radiation by liquid donors, i.e., it evidently increased the total volume of the drops.

The last two effects may be due to an increase in the average density of nonequilibrium carriers (at a constant generation rate) because of a reduction in the diffusion length of excitons in doped germanium [70]. In fact, the scattering of excitons by neutral impurity atoms at liquid helium temperatures may be important even at impurity concentrations $N_i \sim 10^{14}$ cm^{-3}. The probability of such scattering is obviously proportional to N_i and the exciton diffusion coefficient is $D_{ex} \propto N_i^{-1}$. The exciton lifetime in the presence of a moderate number of shallow neutral centers may not change because the recombination process is usually governed by the deep centers. Consequently, the exciton diffusion length $L_{ex} = (D_{ex}\ \tau_{ex})^{1/2}$ in doped germanium should decrease in accordance with $L_{ex} \propto N_i^{-1/2}$. If the depth of the surface layer in in which the exciting radiation is absorbed is much less than the diffusion length and this length does not exceed the thickness of a sample, the average density of nonequilibrium carriers in the active part of a crystal is proportional to $\bar{n} \propto L_{ex}^{-1} \propto N_i^{1/2}$. Obviously, an increase in the average carrier density in the gaseous phase should shift the equilibrium in the two-phase system (kept at a constant temperature) in the direction of the higher total volume of the liquid phase. For this reason, the threshold temperature of the appearance of the condensate drops in doped samples should be higher, for a given excitation rate, than in the case of pure germanium.

Another possible cause of the increase in the resonance absorption and of the threshold temperature in the case of doped samples may be an increase in the concentration of the condensation centers in doped germanium [7]. The impurity atoms and various defects which

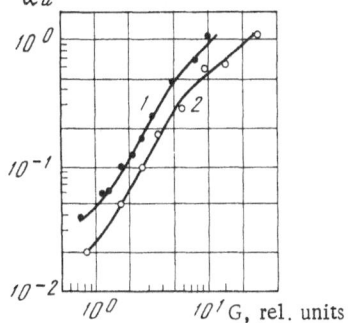

Fig. 20. Dependences of αd on the excitation rate of antimony-doped germanium ($N_{Sb} = 3.6 \times 10^{15}$ cm^{-3}) determined at 1.6°K: 1) $\lambda = 145 \; \mu$; 2) $\lambda = 190 \; \mu$.

appear during doping may act as nuclei of the condensate drops. Obviously, in the presence of a large number of condensation centers, the deepest centers are filled first with drops and this should establish favorable conditions for the appearance of the condensed phase: for example, it may raise the temperature threshold of the condensation and, for a given temperature and generation rate, it may increase the total number of condensed carriers.

Some differences between the behavior of the resonance absorption in doped and fairly pure germanium samples were observed in a study of the dependence of the absorption on the excitation rate. Figure 20 shows such a dependence obtained at T = 1.6°K in the region of the absorption maximum and in the long-wavelength wing of the absorption spectrum of a germanium sample doped with antimony ($N_{Sb} = 3.6 \times 10^{15}$ cm^{-3}). The resonance absorption in the doped sample appeared at lower excitation rates than in intrinsic germanium and when the excitation rate was constant, the absorption was stronger in the doped sample. The nature of the dependence of the absorption on the excitation rate differed considerably for the doped and undoped samples: the initial stage was no longer cubic but a weaker rise of the absorption $\alpha d \propto G^{1.6}$ and then, as in the case of intrinsic samples, the dependence became nearly linear: $\alpha d \propto G^{0.8}$. There was also some tendency for the dependence $\alpha d(G)$ to weaken for the doped samples at the lowest generation rates, particularly in the case of measurements near the resonance spectrum maximum ($\lambda \approx 150 \; \mu$), whereas, in the case of intrinsic germanium, the absorption found in this region decreased rapidly to very low values.

The enhancement of the resonance absorption in doped germanium, compared with the absorption of pure samples at the same excitation levels, and the reduction in the carrier generation threshold can be explained by the two factors considered above: one of these factors is the rise of the average concentration because of the reduction of the diffusion length of excitons and the other is the increase in the concentration of the condensation centers as a result of doping.

To understand the possible reasons for the discrepancies between the dependences on the excitation rate, it would be interesting to compare our results with those obtained in similar investigation of the recombination radiation of condensed drops [70]. In this investigation, the dependence of the intensity of the luminescence of drops in doped germanium was found to be $I_{lum} \propto G^{1.4}$ at a temperature of 1.35°K. These results agreed well with ours in this range. However, at the lowest generation rates near the threshold of the appearance of condensate drops, the dependence of the luminescence intensity on the excitation rate found in [70] was very strong (nearly exponential).

Since the number of condensation centers can be considerably higher in doped samples, the increase in the number of drops f with degree of occupancy of the deepest of the condensation centers should occur in a wide range of carrier generation rates near the threshold. This assumption makes it possible to analyze the kinetics of the condensation process [Eq. (31) in the case of variable values of f] and explain the very rapid rise (including exponential growth) of the total volume of the condensed phase during the initial stage and the subsequent slowing down of this rise. Considerations of this kind are used in the explanation of the experimental results reported in [70].

However, in the resonance absorption measurements, the initial region of rapid rise may be masked by other absorption mechanisms which occur in doped germanium, such as the absorption by self-compensated impurity centers which have captured nonequilibrium photocarriers. This "compensation effect" should obviously contribute to the total absorption in the case when the background illumination corresponds to the absorption band of shallow impurity centers ($\lambda \sim 150\ \mu$), i.e., in the region of the resonance maximum.

The absorption by compensated impurity centers in germanium excited by light causing interband transitions can be estimated experimentally. For example, in the case of germanium doped with antimony ($N_{Sb} = 3.6 \times 10^{15}$ cm^{-3}), the concentration of the compensating centers amounts to $N_A = 7.2 \times 10^{13}$ cm^{-3}. The maximum possible magnitude of the compensation effect can be estimated from the "dark" absorption spectrum of germanium of the same thickness doped with antimony ($N_{Sb} = 2N_A = 1.4 \times 10^{14}$ cm^{-3}) because the intensity and spectral positions of the shallow-donor and acceptor absorption lines are quite close in the case of germanium. This "dark" spectrum, obtained in the absence of excitation of T = 1.5°K, is represented by the dashed curve in Fig. 21. We can see that the compensation effect may play an important role in a fairly narrow part of the spectrum and only in the case of doped and sufficiently strongly compensated samples. However, at low excitation rates, this absorption mechanism may predominate.

We have discussed the results of investigations of the resonance absorption in doped germanium with relatively low impurity concentrations $N_i \leq 3 \times 10^{15}$ cm^{-3}. At higher impurity concentrations, the absorption spectra may be strongly deformed, particularly at long wavelengths (Fig. 21). It is clear from Fig. 21 that the long-wavelength absorption increases considerably with rising impurity concentration. Moreover, the maximum of the spectral curve shifts toward longer wavelengths with rising impurity concentration.

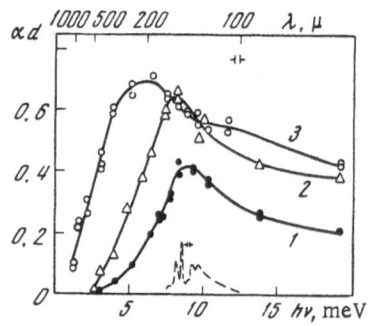

Fig. 21. Spectral dependences of αd determined at 1.5°K and 230 mW for different samples: 1) pure germanium with $N_i \approx 1 \times 10^{12}$ cm^{-3}; 2) antimony-doped germanium with $N_{Sb} = 3.6 \times 10^{15}$ cm^{-3} and $N_A = 7.2 \times 10^{13}$ cm^{-3}; 3) arsenic-doped germanium with $N_{As} = 7 \times 10^{15}$ cm^{-3} and $N_A = 10^{14}$ cm^{-3}; the dashed curve represents the "dark" absorption spectrum.

These changes in the absorption spectrum may be attributed to the influence of impurities on the properties of electron–hole drops, particularly, to the influence of impurity atoms on the damping of plasma oscillations. It is also possible that an increase in the long-wavelength absorption, for example, the increase at $h\nu \sim 5$ meV in the case of arsenic-doped germanium ($N_{As} = 7 \times 10^{15}$ cm^{-3}) in Fig. 21, is due to a basically different absorption mechanism, namely, the absorption of radiation by excitons bound to neutral impurity centers. Further studies would be needed to determine whether this explanation is correct.

CHAPTER IV

RESONANCE LUMINESCENCE OF CONDENSED EXCITON PHASE IN GERMANIUM

§ 1. Experimental Investigation of Resonance Luminescence

Investigations of far infrared properties of strongly excited germanium revealed, apart from the resonance absorption described in the preceding sections, a long-wavelength infrared resonance luminescence [44]. Like the resonance absorption, the luminescence appeared at temperatures below a certain threshold, which depended on the excitation rate. The integrated intensity of this infrared luminescence ($\lambda > 80$ μ) was very high so that we could study in detail its properties, particularly the spectral composition and the dependences on the temperature and excitation rate [64, 65].

Considerable experimental difficulties, specific to the long-wavelength infrared part of the spectrum, were encountered in spectroscopic studies. The low photon energy, compared with the visible or near infrared range, meant that the power of the beam was low for a given photon flux. In spectroscopic measurements the intensity of the luminescence per unit frequency interval (~ 1 cm^{-1}) needed to ensure a resonable spectral resolution, was found to be very low in spite of the considerable intensity of the integrated radiation. In fact, as reported later, we found that the power of the resonance luminescence emerging from a sample was $\sim 10^{-9}$ W per $\Delta\nu = 1$ cm^{-1}. Unfortunately, only a small fraction of this radiation reached a long-wavelength radiation detector. This point can be understood better by considering the method used in spectroscopic measurements.

All the wavelengths emitted by a sample, which was placed in the light-guide cryostat described earlier, passed through an evacuated bent light guide to the monochromator of the DIKS laboratory spectrometer. The dispersive element in the monochromator was an interchangeable echelette diffraction grating. The absence of a preliminary monochromator required careful separation of the radiation with a given wavelength λ from the radiation of higher order ($\lambda/2$, $\lambda/3$, etc.) because the echelette reflected the investigated radiation and all its higher orders in the same direction. The spectrally resolved and filtered radiation reached a bolometer placed in a second light-guide cryostat at the exit from the monochromator. Obviously, the radiation losses in the filtration and in the optical path between the sample and the detector (particularly in the bent light guide) meant that only a small proportion of the energy emitted by the sample reached the detector.

The high threshold sensitivity of the germanium bolometers specially developed for this purpose ($\sim 5 \times 10^{-12}$ W) made it possible to study the spectral composition of the long-wavelength radiation. In the least favorable cases the precision and reliability of the experimental results were increased by repeated measurements and a statistical analysis of the results.

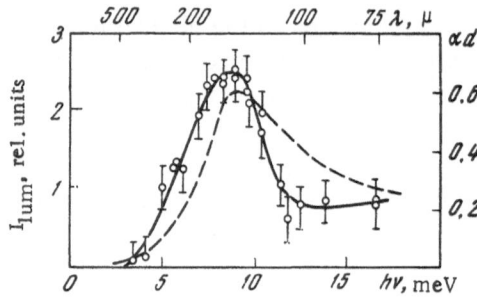

Fig. 22. Spectral composition of the resonance luminescence of germanium at 1.6°K
obtained for an excitation rate of 360 mW
(G = 14). The dashed curve represents the
resonance absorption spectrum.

Measurements of the long-wavelength infrared absorption and luminescence were carried out using the same samples under similar experimental conditions by the method described
above. The luminescence was studied using a differential method in which the exciting radiation was modulated. The same differential method could not be used in the transmission
(absorption) measurements because of the existence of a strong wide-band long-wavelength
luminescence which was much stronger than the probing radiation emerging from the monochromator and used in the transmission measurements.

The far infrared luminescence spectrum of intrinsic germanium, obtained at 1.6°K as
a result of 360 mW excitation, is shown in Fig. 22. At each point the luminescence intensity
was determined allowing for the transmission of the filters and for the distribution function
of the echelette grating; this intensity was reduced to the frequency interval $\Delta\nu = 1$ cm^{-1}.
The dashed curve in Fig. 22 is the resonance absorption spectrum obtained for the same sample
under similar conditions. It is clear from Fig. 22 that the long-wavelength luminescence and
absorption spectra were very similar. The luminescence band was somewhat narrower and
it was shifted in the direction of longer wavelengths. Some differences between the spectra
were also observed in the short-wavelength wing.

The nature of the resonance luminescence was determined by investigating the dependences of its intensity on the temperature and carrier generation rate. As in the resonance
absorption case, it was found that the threshold temperature and the threshold excitation needed
for the appearance of the resonance luminescence were related. These conclusions were drawn
from a study of the luminescence spectrum in a wide photon energy range $h\nu \leq 15$ meV. The
short-wavelength part of the spectrum ($h\nu > 15$ meV) was suppressed by specially selected
filters. The sensitivity of the measurements was considerably higher than in the transmission
(absorption) studies.

The temperature dependences of the integrated luminescence intensity, obtained for
three different rates of excitation of the same sample of intrinsic germanium, are plotted
in Fig. 23. At low excitation rates (curve 1) the luminescence intensity decreased with rising
temperature to a certain constant value. At high excitation rates the luminescence intensity
also fell with rising temperature. However, on transition of liquid helium from the superfluid

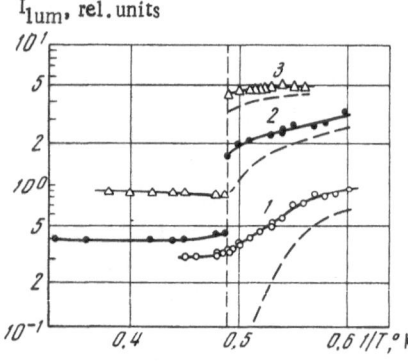

Fig. 23. Temperature dependences of the integrated luminescence intensity ($\lambda > 80$ μ) emitted by germanium subjected to excitation at different rates: 1) 50 mW (G = 3); 2) 120 mW (G =
6); 3) 300 mW (G = 12). The dashed curves represent the same dependences after the subtrac-
tion of the constant background.

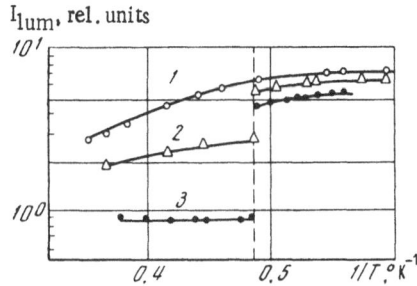

Fig. 24. Temperature dependences
of the integrated luminescence intensity in the region of the λ point
of liquid helium obtained for different germanium samples subjected
to $I_{exc} = 300$ mW: 1) sample doped
with antimony ($N_{Sb} = 3.6 \times 10^{15}$ cm^{-3});
2) sample doped with antimony ($N_{Sb} = 4 \times 10^{14}$ cm^{-3}); 3) pure sample ($N_i \approx 1 \times 10^{12}$ cm^{-3}).

to the normal state, the luminescence intensity fell suddenly severalfold and remained practically constant when the temperature was increased right up to 4°K (curves 2 and 3).

This sudden fall of the luminescence intensity was evidently associated with some overheating of the sample by the exciting radiation which occurred when the heat exchange between
the sample and the cooling agent deteriorated as a result of the transition of liquid helium to
the normal state. This sudden change in the luminescence intensity was considerably smaller
in the case of germanium doped with shallow impurities. The discontinuity decreased with
rising impurity concentration and when the concentration of antimony reached $N_{Sb} = 3.6 \times 10^{15}$
cm^{-3}, the transition across the λ point of liquid helium was practically smooth (Fig. 24).

It was reported in Chap. III that the temperature threshold of the resonance absorption
of doped germanium was shifted in the direction of higher temperatures, compared with intrinsic germanium. A similar behavior was exhibited by the threshold temperature of the
resonance luminescence. The results obtained for doped germanium confirmed that the overheating of the sample due to the transition at the λ point could not be large. In any case, the
temperature of the overheated sample did not exceed 3–4°K, which was the threshold temperature for the rates of excitation of germanium samples doped with shallow impurities in
concentrations of 10^{15} cm^{-3}. However, the temperature dependences of the resonance absorption and luminescence obtained at low excitation rates for intrinsic germanium indicated that
even a small change in the temperature near the threshold was sufficient for the disappearance
of these resonance phenomena.

When measurements were made in superfluid helium a sample was not overheated even
at the highest powers of the exciting radiation used in our study. The thermal conditions were
monitored by the same methods as those described in §3, Chap. II.

The fraction of the integrated radiation weakly dependent on temperature was still observed at liquid helium temperatures above the λ point up to 4.2°K (Fig. 23). This fraction
corresponded to the photon energy range $h\nu > 11$ meV in the spectrum of Fig. 22. It represented about one-eighth of all the integrated luminescence, which was in agreement with a
sevenfold or eightfold reduction in the luminescence intensity observed when the temperature
was increased from 1.6 to 4.2°K (Fig. 23). It was natural to assume that the luminescence
which was observed above the threshold and which depended weakly on the temperature was
different from the main resonance luminescence. Thus, by subtracting the constant background

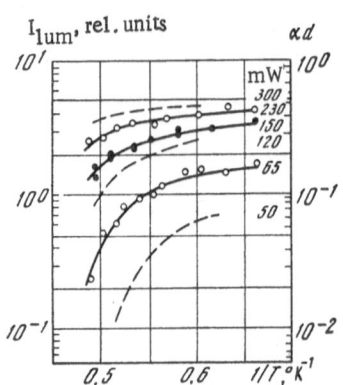

Fig. 25. Temperature dependences of the resonance
absorption (continuous curves) and resonance luminescence (dashed curves), plotted for intrinsic germanium
subjected to different excitation rates given alongside
the curves.

from curves 1–3 in Fig. 23 we could obtain the corrected temperature dependences of the
"pure" resonance luminescence. These corrected dependences are represented by dashed
curves in Fig. 23 and they are compared in Fig. 25 with the temperature dependences of the
resonance absorption under similar experimental conditions.

It is clear from Fig. 25 that the temperature dependences of the resonance absorption
of condensed electron–hole drops in germanium were practically identical with the corresponding temperature dependences of the resonance luminescence of germanium observed under
similar experimental conditions. In view of the similarity of the absorption and luminescence
spectra mentioned above and the similarity of the temperature dependences, we concluded that
the far infrared resonance luminescence of germanium observed at low temperatures was due
to the condensation of excitons into electron–hole drops.

An even greater similarity of the resonance absorption and luminescence was manifested
at 1.5–1.6°K when a comparison was made of their dependences on the excitation rate. The
dependences obtained for the integrated luminescence at two temperatures (1.6 and 4.2°K) are
shown in Fig. 26. Curve 1, obtained at 1.6°K, exhibited – as in the absorption case – a region
of rapid rise of the luminescence intensity $I_{lum} \propto G^{2.7}$, which eventually changed to an almost
linear dependence. Moreover, the much higher sensitivity in the integrated luminescence measurements allowed us to record a flatter region at very low excitation rates. We assumed that
the luminescence which depended weakly on the excitation rate and was observed at low (subthreshold) values of this rate, differed in its nature from the resonance luminescence. In this
way we could divide the curve into two components so as to separate the "pure" resonance
luminescence from the combined curve. This procedure increased the exponent in the power-
law dependence of the resonance luminescence intensity on the excitation rate: in the rapidly
rising region this exponent increased from 2.7 to 3.2 (dashed curves in Fig. 26).

At 4.2°K the integrated luminescence depended relatively weakly on the excitation rate
practically throughout the investigated range. A somewhat faster rise was observed only at

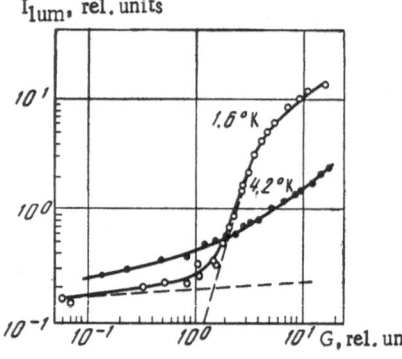

Fig. 26. Dependences of the integrated luminescence
intensity ($\lambda > 80 \mu$) on the rate of excitation of germanium at two temperatures.

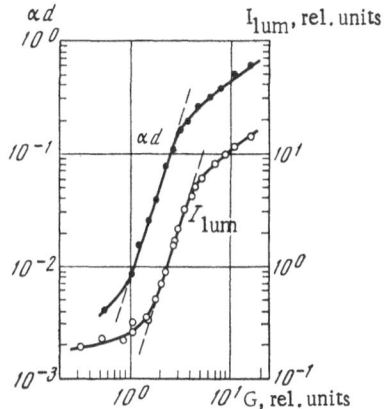

Fig. 27. Dependences of the resonance absorption and resonance luminescence of intrinsic germanium on the rate of excitation at 1.6°K.

the highest excitation rates. It was likely that at these rates the resonance luminescence was observed even at 4.2°K but it just exceeded the background of the slowly varying subthreshold luminescence.

Thus, we could assume that the resonance luminescence obeyed, like the resonance absorption, an approximately cubic dependence of the intensity on the excitation rate, which then became linear. The similarity of the dependences in the two effects was due to the appearance of condensed electron–hole drops in germanium and they manifested the kinetics of the general rise of the volume of the condensate, discussed in Chap. III. Figure 27 shows clearly the great similarity of the resonance absorption and luminescence curves.

§2. Discussion of Experimental Results. Effective Luminescence Temperature of Drops

In the preceding section we demonstrated that the far infrared resonance luminescence of germanium was very similar to the resonance absorption. The two phenomena occurred at temperatures below a certain threshold, which depended on the excitation rate; they had similar dependences on the excitation rate and similar spectra. These experimental observations indicated that the two effects had the same origin, i.e., both of them could be attributed to the appearance of condensed electron–hole drops in germanium at low temperatures.

The resonance absorption was interpreted in preceding chapters by the dipole model of the absorption of electromagnetic energy by electron–hole drops in the plasma oscillation region. In view of the common origin of the resonance absorption and luminescence, we interpreted the latter as the dipole luminescence of electron–hole drops due to the excitation of plasma oscillations in these drops. We shall now apply this model to the experimental data on the resonance luminescence and in this way we shall supplement our information on the properties of electron–hole drops.

If the resonance luminescence of electron–hole drops is of equilibrium nature, the luminescence and absorption of drops in the same part of the spectrum should be related by the Kirchhoff law, which is a consequence of a more general thermodynamic principle of detailed equilibrium. In order to apply this principle, we must be sure that an equilibrium distribution of the particle energies exists in a drop. Then, the particles in a drop which are in equilibrium with one another cannot be in equilibrium with the surrounding medium, i.e., with the crystal lattice. This means that the particles in a drop have their own temperature, which differs from the lattice temperature. This situation may occur if the time for the establishment of an equilibrium between particles in a drop is sufficiently short compared with the time for the establishment of an equilibrium between the drop and the lattice.

It is pointed out in Chap. III that the interaction of electrons and holes in a drop with the crystal lattice may be weakened considerably because of the Fermi degeneracy of carriers. In the case of an equilibrium particle concentration in a drop $n_0 = 2 \times 10^{17}$ cm^{-3} the mean free path of carriers colliding with phonons at temperatures 2–4°K is of the order of 10^{-6}–10^{-7} sec [42]. Electron–hole collisions may play an important role in drops with such a high particle concentration. The results of electrical measurements [37] can be used to estimate the average time between electron–hole collisions and for the stated particle concentration this time is 10^{-10} sec at 1.7°K. Under these conditions we can regard electron–hole drops as quasiequilibrium systems whose temperature may differ from the lattice temperature.

It follows from the Kirchhoff law that the ratio of the emissivity of a body to its absorptivity is a universal function of the emission frequency and of the temperature of the emitting body. Consequently, the results of measurements of the resonance absorption and luminescence of electron–hole drops in germanium, carried out in the same part of the spectrum, allow us to use the Kirchhoff law in an estimate of the effective temperature of drops on the assumption that the resonance luminescence is an equilibrium effect.

An estimate of the effective temperature of drops was made on the basis of the experimental data in the region of the maxima of the absorption and luminescence spectra. In these calculations we needed to know the absorption coefficient and the absolute luminescence power. The absorption at the maximum was $\alpha d = 0.6$ when the excitation rate was 360 mW and the temperature was 1.6°K. The absolute luminescence power under the same conditions was determined allowing for the transmission of the filters, distribution in the echelette, and losses in the light-guide system. This was done in two ways: 1) by comparing the luminescence signal with the bolometer noise, which was measured separately; 2) by comparing the observed luminescence with the intensity of the long-wavelength infrared radiation emitted by a mercury lamp [74]. Estimates deduced by both methods gave similar values of the spectral density of the luminescence power $I_\nu \sim 10^{-9}$ W/cm^2 in a frequency interval $\Delta\nu = 1$ cm^{-1}. The temperature of electron–hole drops was determined using the standard expression for the radiation emitted by a sample, subject to an allowance for the internal reflection and self-absorption [75]:

$$I_\nu = \frac{(1-R)\,[1-\exp(-\alpha d)]}{1 - R \exp(-\alpha d)}\, Q(T^0)\,\frac{d\Omega}{\pi}, \tag{35}$$

where R is the reflection coefficient of germanium; $d\Omega$ is a solid angle in which the radiation is emitted; $Q(T^0)$ is the Planck function defined by

$$Q(T^0) = \frac{2\pi}{h^2 c}\, \frac{(h\nu)^3}{\exp\left(\dfrac{h\nu}{kT}\right) - 1}. \tag{36}$$

The effective temperature of electron–hole drops found in this way for the conditions described above was $T_d \approx 15$°K. The solid angle used in these calculations was π, because the luminescence emitted by a sample was largely integrated over the angles in the light-guide system. It should be pointed out that since the drop temperature occurred in the argument of the exponential function in Eq. (36), the error in the estimate of the absolute luminescence power by even an order of magnitude could not alter the calculated value of T_d by a factor exceeding 1.5.

The temperature of electron–hole drops could also be estimate by a different method involving an analysis of the luminescence spectrum and a comparison with the absorption spectrum. Since in the equilibrium case the absorption and luminescence spectra are related by Eq. (35), the temperature of the luminescing objects could be found by plotting the frequency dependence of the ratio of the spectral density of the luminescence to the magnitude

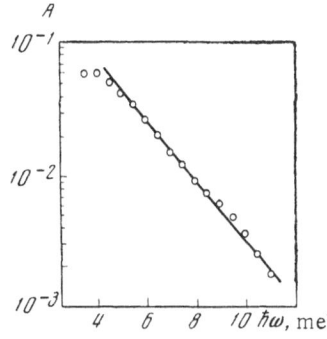

Fig. 28. Spectral dependence A = [I /(hν)³]·[1 − R exp (− αd)]/[1 − exp(− αd)], plotted using the experimental results obtained at 1.6°K for I_{exc} = 360 mW.

of the absorption:

$$\ln\left[\frac{I_\nu(\nu)}{(h\nu)^3}\frac{1-Re^{-\alpha(\nu)d}}{1-e^{-\alpha(\nu)d}}\right]=\varphi(\nu). \tag{37}$$

According to Eqs. (35) and (36), in the case of the equilibrium luminescence the dependence $\varphi(\nu)$ at frequencies such that $\exp(h\nu/kT) \gg 1$ should be linear with a slope governed by the temperature of the luminescing objects:

$$\varphi(\nu)=\mathrm{const}-\frac{h\nu}{kT}. \tag{38}$$

The experimentally obtained luminescence $I_\nu(\nu)$ and absorption $\alpha(\nu)$d spectra were used to plot the dependence of the logarithm on the left-hand side of Eq. (37) on the luminescence frequency (Fig. 28). It is clear from this figure that the experimental points fitted well a single straight line whose slope corresponded to $T_d \approx 20°$K.

Figure 29 shows the equilibrium luminescence spectra calculated on the basis of the Kirchhoff law and the experimental absorption spectra at several values of the drop temperature. A satisfactory agreement between the calculated and experimental luminescence spectra was observed at $T_d = 20°$K, whereas at lower and higher temperatures there was a considerable discrepancy between the calculated and experimental curves.

Thus, an estimate of the effective temperature of electron−hole drops from the absorption and luminescence spectra, obtained on the assumption that the luminescence was an equilibrium effect, gave $T_d \sim 15$-$20°$K, which was considerably higher than the lattice temperature under the experimental conditions.

The difference between the drop temperature and the lattice temperature is not surprising. In fact, as mentioned earlier, the weak coupling between carriers in the condensed phase and the lattice may result in the establishment of a quasiequilibrium state in a drop and this state can have its own temperature. An overheating of a drop may result from, for example, the

Fig. 29. Comparison of the resonance luminescence spectra determined experimentally at 1.6°K (points) and calculated from the absorption spectra (curves) for different values of the effective temperature of electron−hole drops: 1) $T_d = 10°$K; 2) $T_d = 20°$K; 3) $T_d = 25°$K.

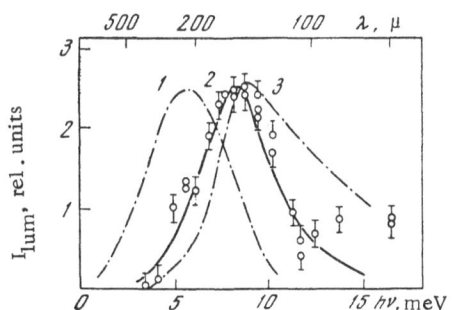

capture of carriers from the gaseous phase during the growth of the drop or it may be due
to steady-state recombination of carriers in the drop. In the former case some of the energy
liberated as a result of the transition of carriers to the condensate may be transferred to the
particles in a drop. In the latter case the energy may be transferred to a drop in several
ways. For example, part of the energy of carriers recombining to produce a photon may be
transferred directly to electrons and holes in a drop [43]. Moreover, in the case of radiative
recombination of carriers forming a condensate some energy is evolved in the form of monochromatic phonons, mainly of the LA, TO, and TA type [34, 43]. A fairly large energy is also
liberated in a drop in the case of nonradiative carrier recombination.

Thus, local energy evolution processes may occur in a drop and some of the evolved
energy may be transferred in one way or the other to the particles in the drop. When, a
fraction of this energy is transformed into the energy of thermal motion of electrons and
holes in a drop, the latter may become hotter than the crystal lattice. However, the drop
temperature $T_d \sim 15$-$20°K$ is almost twice as high as the critical temperature of the condensed phase, i.e., the drop evaporation temperature, estimated from the experimental data
in [43, 53, 70]. This raises doubts about the equilibrium nature of the luminescence of electron–hole drops and about the value of the drop temperature T_d obtained on the basis of this
assumption.

Clearly, the resonance luminescence of electron–hole drops appears as a result of excitation of plasma oscillations which is a basically nonequilibrium process in which the energy
transferred to the particles in a drop is largely transformed directly into the energy of plasma
oscillations and not into the thermal energy. This means that when the energy is supplied to
a drop we may find that under certain conditions the process of excitation of plasma oscillations,
i.e., of the collective motion of electrons and holes in a drop, may be stronger than the increase
in the kinetic energy of the individual carriers, i.e., it may be stronger than the carrier heating.

In the case of nonequilibrium excitation of the luminescence of electron–hole drops the
luminescence intensity may be considerably higher than the intensity of the equilibrium (thermal)
luminescence. In this case the above value of T_d does not represent the thermal motion of
individual electrons and holes in a drop but the degree of disequilibrium of the luminescence
process. Therefore, the value $T_d \sim 15$-$20°K$ should, more correctly, be called the effective
luminescence temperature of drops.

§3. Influence of Inhomogeneous Deformation on

Resonance Absorption and Luminescence. Mobility of

Electron–Hole Drops

According to the hypothesis put forward in [10, 42], one of the most characteristic features
of electron–hole drops is their extremely high mobility because the density of effective masses
in a drop is very low: $m_d \sim m_{ex}/a_{ex} \sim 10^{-10}$ g/cm³. Electron–hole drops may be accelerated
by inhomogeneous deformation, inhomogeneous magnetic fields, etc., which may displace drops
over considerable distances.

The problem of the motion of drops in an inhomogeneous strain field $\varepsilon_{ik}(\mathbf{r})$ is considered
in [10], where the following expression is obtained for the velocity of a drop in a field $\varepsilon_{ik}(\mathbf{r})$:

$$\mathbf{v} = \frac{D_{ik}\tau_v}{m_{ex}} \operatorname{grad} \varepsilon_{ik}, \tag{39}$$

where D_{ik} is the combined deformation potential of an electron and a hole; τ_v is the effective
relaxation time of the drop momentum. In pure semiconductors the time τ_v is governed by
collisions with phonons and, because of the Fermi degeneracy of carriers in a drop, this time may

be fairly long: according to an estimate given in [42], $\tau_v \sim 10^{-6}$ sec for drops in germanium at $T \approx 2°K$. The more frequent collisions of electrons and holes with one another inside a drop do not alter the momentum of the drop as a whole but maintain the Fermi distribution of the particle energies in a drop (in a coordinate system moving together with the drop) but the temperature of the drop may be different from the lattice temperature.

The application of Eq. (39) to germanium shows [43] that even for a small pressure gradient, of the order of 10-100 kg/cm^2 per 1 cm, electron—hole drops may be accelerated to the velocity of sound. The subsequent rise of the drop velocity is limited because at supersonic velocities a drop considered as a whole may radiate acoustic waves and this will alter the translational velocity. Drops may travel fairly long distances in a crystal when they are accelerated by a inhomogeneous strain field to velocities of the order of the velocity of sound.

We attempted to detect experimentally the motion of drops by measuring simultaneously the resonance absorption and luminescence under inhomogeneous deformation conditions. In these measurements we produced a fairly strong local deformation by applying a steel needle to the edge of a fairly long (l = 25 mm) sample of pure germanium. The measurements of the resonance absorption at the maximum of the spectrum and of the integrated luminescence were carried out in two parts of the sample which were separated by very different distances from the strong-deformation region. The separation from the needle to the measurements zone, i.e., to the center of an exciting light spot of 9 mm diameter, was 5.5 mm in one case and 16.6 mm in the other. The resonance absorption in the region adjoining directly the needle decreased by a factor of 4 compared with the absorption in the more distant region which was hardly deformed.

Similarly, the integrated luminescence intensity (obtained at the same excitation rate) measured close to the needle was approximately 4 times lower than the intensity of the luminescence measured far from the needle. The change in the dependence of the integrated luminescence intensity on the excitation rate in a region close to the needle was of the type shown in Fig. 30. We found that the whole dependence $I_{lum} = f(G)$ and its threshold shifted strongly in the direction of higher excitation rates when the measurements were carried out near the source of strong deformation.

We could assume that the nonequilibrium lifetime of carriers excited at distances of several millimeters from the source of local deformation did not change greatly as a result of this deformation. This assumption was confirmed by a study of the free-exciton luminescence emitted by germanium under pressure [43], which showed that the intensity of the exciton luminescence line was hardly affected even in the case of fairly strong homogeneous and inhomogeneous deformations. In this investigation the carrier lifetime in a drop was measured in a crystal subjected to a uniaxial compression: the application of pressures of the order of 400 kg/cm^2 reduced the lifetime in a drop by a factor of just 1.5.

Thus, if it was assumed that the lifetimes of electrons and holes were not affected by a local deformation at distances of several millimeters from the region where these carriers

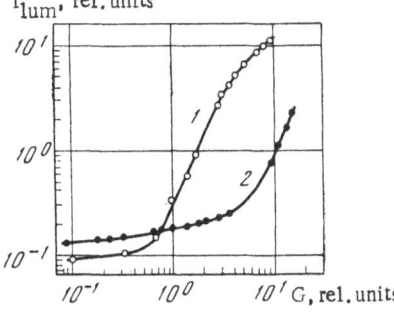

Fig. 30. Dependences of the integrated luminescence intensity of germanium on the rate of excitation at 1.6°K under inhomogeneous deformation conditions. Curves 1 and 2 represent the results obtained for an excitation region located at a distance of 1.5 and 5.5 mm, respectively, from the needle which generated the deformation.

were excited, the reduction in the resonance absorption and luminescence in the region close to a needle could be explained only by the motion of electron—hole drops in the strain gradient, i.e., their escape from the region being examined. The experimental results could be used to estimate approximately the distance traveled by drops in the deformation field during their lifetime, which was approximately 1.5 cm, the value of which was much greater than the diffusion length of free carriers and excitons.

Experimental observations supporting the hypothesis of extremely high mobility of electron—hole drops were reported also elsewhere [42, 43]. An extremely strong fall of the intensity of the drop luminescence line was observed in germanium under inhomogeneous deformation conditions and the maximum of the line shifted much faster than expected for an average stress field in a sample. These observations were attributed to the escape of electron—hole drops from the region of maximum deformation.

CHAPTER V

PHOTOIONIZATION AND EXCITATION OF FREE EXCITONS IN GERMANIUM BY SUBMILLIMETER RADIATION

§ 1. Photoionization and Excitation Spectra

Investigations of the far infrared absorption spectra of excited germanium revealed for the first time the photoionization of free excitons in semiconductors by submillimeter radiation [76]. The measurements were carried out on selected samples of undoped germanium which had sufficiently long indirect-exciton lifetimes, as judged by the intensity of the free-exciton luminescence line (714.2 meV). The exciton lifetime in these samples reached several microseconds at liquid helium temperatures and the residual impurity concentration did not exceed 1×10^{14} cm^{-3}. The method of excitation of samples of about 1 cm^2 area and 0.5 mm thick and the spectroscopic measurement technique were both described in Chap. II. The quantity measured directly was the same as in studies of resonance absorption (Chap. III): we measured the ratio of the transmission of an excited sample to its "dark" transmission, which was then transformed into the product αd representing the induced optical absorption.

The first attempts to observe photoionization of free excitons in germanium at 1.5°K were unsuccessful. The reasons for this became clear later. At this temperature and for the excitation rates employed the majority of excitons condensed into a new phase of electron—hole drops [44, 64, 65] (see Chap. III) and the number of excitons remaining in the gaseous phase was insufficient to produce a significant absorption of long-wavelength infrared radiation. An investigation of the properties of the condensed phase in germanium not only gave results of intrinsic interest but made it possible to approach anew the problem of observation of photoionization of free excitons. It became clear that in order to study free excitons one would have to carry out measurements in a different part of the concentration—temperature phase diagram in which the equilibrium would be shifted in the direction of the exciton gas. This could be done by increasing the temperature of a crystal.

According to the experimental results obtained in the present investigation (Chap. III), the upper temperature limit (threshold) of existence of the condensed phase in undoped germanium ($N_i < 1 \times 10^{14}$ cm^{-3}) was 3.0-3.5°K at the maximum excitation rates which could be achieved with our sources (Fig. 15). When the temperature of a sample is raised above this threshold, the number of excitons in the gaseous phase should increase considerably because of the evaporation of the condensed drops.

The intensity of the luminescence representing the free exciton luminescence rose strongly in the region of the temperature threshold of the luminescence line representing the condensed

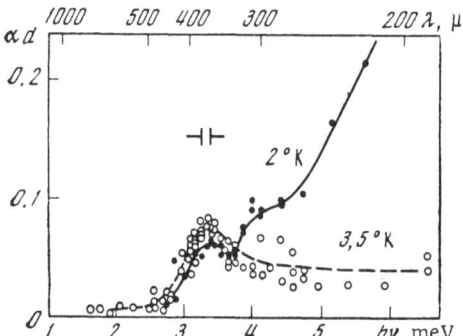

Fig. 31. Spectral dependence of αd determined at various temperatures for I_{exc} = 300 mW.

exciton phase in germanium (Fig. 15) and this was evidently due to an increase in the number of free excitons [43]. The concentration of free excitons in the gaseous phase could increase also on further rise of temperature because of the reduction in the probability of the capture of excitons by nonradiative recombination centers and because of an increase in the lifetime until the thermal dissociation of excitons into free electrons and holes became a significant process. On the basis of these considerations we carried out measurements of the transmission spectra of excited germanium in the far infrared at temperatures in the range T > 2°K.

When the temperature of a sample was raised from 1.5 to 2.0°K, we observed some reduction in the resonance absorption (Fig. 16) and two singularities in the long-wavelength edge at photon energies of 3.4 and 4.0 meV (these singularities were absent from the spectra obtained at 1.5°K). Details of the long-wavelength resonance absorption spectrum obtained at 2.0°K are shown in Fig. 31. A further rise of the temperature reduced the sensitivity of the detector of long-wavelength infrared radiation and raised considerably the noise level. These difficulties, encountered in the measurements of the transmission at temperatures above the λ point of liquid helium, were due to fluctuations of a liquid helium column in the light-guide tube when the sample was strongly illuminated. The reliability of the results obtained under these conditions was increased by repeated measurements of the transmission at 3.5°K. It is clear from Fig. 31 that the absorption due to the electron—hole drops was practically absent from the spectrum in the range $h\nu \geq 5$ meV obtained at 3.5°K and the absorption maximum at 3.4 meV appeared much more clearly than at 2.0°K, in spite of large fluctuations of the background.

In order to retain the high sensitivity of the measurements and yet avoid the exciton concentration, the subsequent measurements were carried out at 1.5°K but a sample was thermally isolated from liquid helium by a thin (~ 40 μ) polyethylene terephthalate (Mylar) film, which was transparent to long-wavelength infrared radiation and to the exciting light. Under these conditions the temperature of the sample rose because of the absorption of the exciting radiation energy. The temperature of the sample was measured by mounting a min-

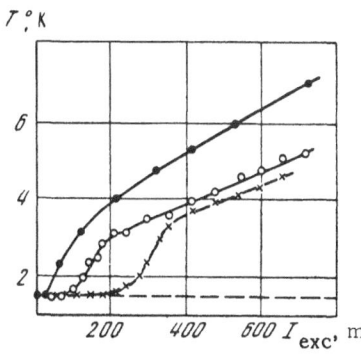

Fig. 32. Dependences of the temperature of a germanium sample on the intensity of excitation for different thermal insulation conditions. The topmost curve corresponds to the most effective thermal insulation.

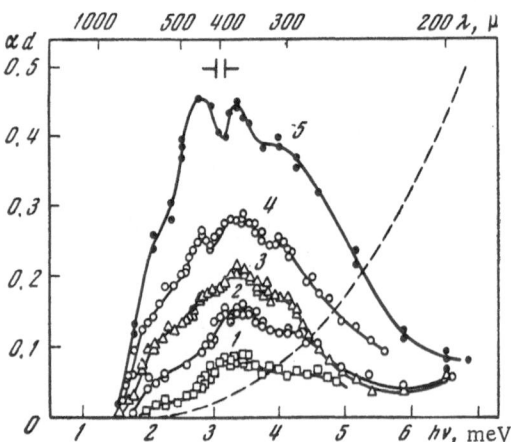

Fig. 33. Spectral dependences of αd determined for a thermally insulated germanium sample at liquid helium temperature of 1.5°K for different excitation rates (mW): 1) 300; 2) 380; 3) 440; 4) 520; 5) 600.

iature carbon microthermometer directly on its surface (the characteristics of this thermometer were given in Chap. II — see Fig. 7).

The dependences of the temperature of the sample on the intensity of the exciting radiation, determined under different thermal insulation conditions, are plotted in Fig. 32. It is clear from this figure that this method made it possible to raise the temperature of the sample to 6-7°K and yet retain the optimal temperature of liquid helium which was 1.5°K.

The absorption spectra of such a thermally insulated germanium sample were measured using different excitation rates, i.e., at different temperatures of the sample (Fig. 33). Curve 1 was obtained for $I_{exc} = 300$ mW when the temperature of the sample was 3.0-3.5K; curve 3 was obtained for 440 mW and 4.0-4.5°K; curve 5 was obtained at the highest excitation rate of 600 mW and the temperature of the sample was then 5-6°K. It was clear (Fig. 33) that under these conditions there was no resonance absorption in the condensed phase, whose long-wavelength edge (measured for the same samples at 1.5°K without overheating at an excitation rate of 520 mW) is represented by the dashed curve. The absorption spectra obtained were in the form of a wide band with two maxima at photon energies of 3.35 and 2.8 meV; the long-wavelength maximum appeared and grew when the temperature was increased. All the curves obtained at different temperatures exhibited also a singularity at the photon energy of 4.0 meV and a definite kink in the long-wavelength edge at 2.1 meV.

Since the observed absorption spectra corresponded to the binding energy of free indirect excitons in germanium and since they were observed only for sufficiently pure samples selected in accordance with the exciton lifetime, it was natural to assume that these spectra represented the interaction of the long-wavelength infrared radiation with free indirect excitons.

The absorption in the observed band depended strongly on the conditions on the surface of a crystal in the case of measurements carried out on thermally insulated samples and on samples which were in contact with liquid helium at temperatures of 2.0-3.5°K. These conditions were not easily controlled and they could alter the surface recombination velocity of excitons, depending on the state of the surface. The spectra shown in Figs. 31 and 33 were obtained under surface state conditions which were close to optimal.

§2. Discussion of Experimental Results. Energy

Levels of Excitons

The theoretical and experimental data on the energy levels of indirect excitons in germanium published in the literature and discussed in Chap. I are confined entirely to the ground state which is split into two levels (Table 2). The complete energy spectrum of excitons in germanium has not yet been calculated and there is no published information on the energies of excited states. Since the available information was insufficient for the interpretation of the observed absorption spectra, the main conclusions were drawn by a comparison with a different and more thoroughly investigated hydrogen-like electron system, i.e., with a system of shallow group V donor impurities in germanium. The similarity of the theoretical models and methods used in the description of these systems (see Chap. I) provided the basis for this comparison. Thus, the theoretical description of the exciton states in germanium, obtained subject to certain simplifying assumptions in relation to the form of the Hamiltonian of the effective mass of holes, was known to be similar to the description of the states of shallow donors subject to the allowance for the appropriate values of the reduced effective mass.

In the case of shallow donors in germanium the theory is in good agreement with the experimental results. The long-wavelength infrared absorption spectra of shallow donors in germanium are characterized by two narrow and strong lines of type B, which are due to transitions from a chemically split (into two levels) ground state to excited levels $(2p, \pm 1)$ (Fig. 34a) [12]. Similar narrow but much weaker lines of type A, C, and E (the last one is not shown in Fig. 34) are due to transitions from the ground to excited states whose quantum numbers are $(3p, \pm 1)$, $(3p, 0)$, and $(2p, 0)$, respectively. The transitions to the conduction band are manifested by a flat (type I) weak maximum. The position and relative intensities of the lines in the experimentally observed spectra are in good agreement with the results of theoretical calculations (Fig. 34) [12, 19]. The agreement between the theory and experiment is particularly good for the excited levels and for the higher ground-state level, whose positions are found to be practically independent of the chemical nature of the impurity centers.

We may expect that basically the structure of the absorption spectrum of free excitons in germanium is similar to that observed for shallow donor centers, with the exception that the width of the lines in the exciton spectrum will be governed by different porcesses and will be considerably greater. Since the separations between the individual levels in the energetically more compact exciton system are smaller than in the case of impurities, the exciton spectrum should differ from the line spectra of impurities and, because of the superposition of neighboring lines, it should have the form of a wide band with some structure. The general appearance of the measured absorption spectra (Fig. 33) is in agreement with these conclusions.

By analogy with the impurity absorption spectra of germanium, the observed structure of the exciton absorption spectrum, namely the absorption maxima at photon energies of 3.35 and 2.8 meV, may be attributed to type B transitions from the two lower states of an exciton to the corresponding excited states. The conclusion that these lines are due to transitions from different initial states is supported by the different temperature dependences of their intensities. The experimentally observed relative increase of the intensity of the absorption line at 2.8 meV with rising temperature is a natural result of the increase in the population of the upper of the two exciton levels. Applying the same analogy and using the similarity of theoretical models of excitons and donor impurities, we can use the position of the type B

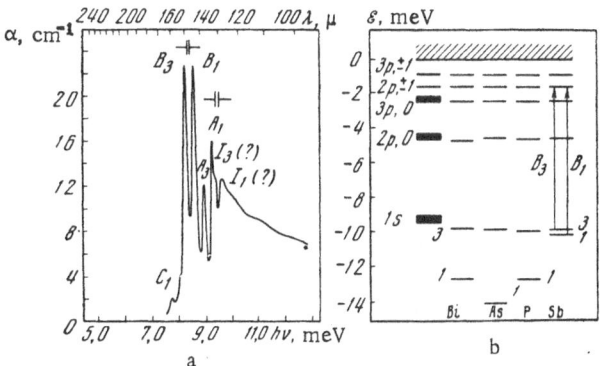

Fig. 34. Impurity absorption spectrum of antimony-doped ($N_{Sb} \approx 1 \times 10^{14}$ cm^{-3}) germanium recorded at T = 9°K (a) and energy-level scheme of donor impurities in germanium (b) taken from [12].

absorption line to find the ionization energy of excitons in germanium in the same way as this is done in the analysis of the absorption spectra of donor impurities [12].

A calculation of this kind carried out for two lower levels of an exciton gives the energies $\mathcal{E}_{ex_1} = 4.0$ meV and $\mathcal{E}_{ex_2} = 3.34$ meV. Similar estimates can be obtained also for other transitions. In particular, in the case of type E transitions the energy is 2.1 meV, which is in agreement with the photon energy corresponding to a pronounced kink in the long-wavelength edge of the absorption spectrum. The energy of an exciton in the lowest state is found to be 4.0 meV, which also corresponds to a singularity found in the absorption spectra and is evidently associated with the photoionization of excitons (Fig. 33).

Thus, the experimentally determined photoionization and excitation spectra of free indirect excitons in germanium can be explained by an energy scheme of the observed optical transitions shown in Fig. 35. According to this scheme, the binding energies of indirect excitons in germanium are 4.0 and 3.34 meV for the two lowest energy states 1 and 2. The error in the determination of the energies of the optical transitions used in the calculation of the energies of levels 1 and 2 does not exceed 0.2 meV. However, the approximate nature of the calculation model which is based on the analogy with the energy spectra of shallow donor impurities in germanium may give rise to somewhat larger errors in the determination of the absolute energies of exciton levels. Each of the two levels (1 and 2) of the split ground state is associated with a system of excited levels whose positions can be found from the energies of the corresponding optical transitions. Estimates show that there are two pairs of excited exciton levels with energies of approximately 0.55 and 0.65 meV, and 1.6 and 1.9 meV.

The splitting of the ground state of indirect excitons in germanium is due to the complex structure of the valence band and due to the anisotropy of the effective electron mass [20, 23]. According to our measurements, the splitting is $\Delta_{ex} = 0.66 \pm 0.1$ meV, which is less than the values 0.8 and 1.0 meV found from the experimental studies of the fundamental absorption edge of germanium [23, 25, 28], but is in good agreement with the value $\Delta_{ex} = 0.6$ meV, found theoretically in [20] (Table 2). The value $\Delta_{ex} = 0.66$ meV, which can be expressed in terms of temperature as 8°K, leads us to expect a considerable temperature dependence of the intensity of the optical transitions from the upper split state at temperatures of 3-6°K, which is indeed observed for the absorption line at 2.8 meV.

A recent differential spectroscopic investigation of the structure of the fundamental absorption edge of germanium [28] has revealed two peaks in the spectrum of the second derivative of the absorption coefficient with respect to the wavelength: these peaks are found in the region corresponding to transitions assisted by a TA phonon and they are separated from one another by 2.4 meV. These peaks are attributed to indirect forbidden transitions to the exciton states with n = 1 and n = 2 (n is the principal quantum number) and the energy separating them is used to find, in the hydrogen-like approximation, the value of the Rydberg constant of indirect excitons in germanium. This approach, justified in [2, 17] for semiconductors with

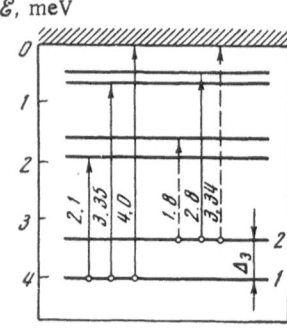

Fig. 35. Optical transitions between energy levels of an indirect exciton in germanium used to explain long-wavelength infrared absorption spectra.

simple spherical energy bands (for example, for cuprous oxide crystals exhibiting a hydrogen-like series of absorption lines [18]), is clearly inapplicable to germanium [77].

The binding energies of indirect excitons in the ground state, 4.0 and 3.34 meV, obtained in the present study from the experimental photoionization and excitation spectra of excitons by comparison with the energy structure of shallow donor impurities in germanium, are higher than the theoretical and experimental values obtained in other investigations (see Table 2). The precision of the determination of the level energies by this method is discussed above. The absolute values of the binding energy of indirect excitons found in experimental investigations of the structure of the fundamental absorption edge of germanium are subject to an indeterminacy which is due to the absence of sufficiently reliable data on the position of the energy-band edges. This may explain the considerable discrepancies between the binding energies obtained by different workers. In relation to the theoretical values of the binding energy of indirect excitons in germanium we must point out that calculations carried out for shallow donor [12, 19] and acceptor [13] impurities in germanium and for indirect excitons in silicon [20, 27] systematically give ground-state binding energies which are smaller than those found experimentally.

We shall now consider briefly the processes which govern the spectral width of the far-infrared exciton absorption lines. In this respect the exciton spectrum should differ considerably from the impurity absorption spectra. It is clear (Fig. 33) that instead of the narrow lines of impurities, the exciton absorption spectrum consists of bands of about 1 meV half-width, which corresponds to 2kT at temperatures of 5-6°K. It should be noted that at the investigated temperatures the mechanism responsible for the width of the impurity absorption lines, namely the interaction between impurity atoms and zero-point vibrations of the crystal lattice, does not play any significant role in the case of excitons because the intensity of this interaction is inversely proportional to the square of the Bohr radius [78]. However, a more important point is that the excitons, which are highly mobile particles, have an intrinsic kinetic energy of the order of kT and they may interact with phonons of similar momentum. This should result in broadening of the exciton lines to $\Delta(h\nu) \sim kT$. Moreover, there may also be a strong interaction between excitons and nonequilibrium long-wavelength phonons which may accumulate in a crystal as a result of thermalization of carriers generated by interband absorption of light.

The experimental data on the absorption due to free excitons can be used to determine the photoionization cross section of free indirect excitons in germanium:

$$\sigma_{ex} = \frac{\alpha d}{N_{ex}}, \tag{40}$$

where N_{ex} is the total number of excitons per unit area of the excited surface. The value of N_{ex} can be found by assuming the exciton lifetime ($\tau_{ex} \sim 10^{-6}$ sec) and finding experimentally — from the room-temperature photoconductivity of germanium — the rate of generation of electron-hole pairs G. The experimental parameters are $G \sim 10^{18}$ cm$^{-2} \cdot$sec^{-1}, $N_{ex} \sim 10^{12}$ cm^{-2}, and $\alpha d \approx 0.1$. Thus, according to the experimental results the photoionization cross section of indirect excitons in germanium is $\sigma_{ex}^{exp} \sim 10^{-13}$ cm^2.

This value may be compared with the theoretical estimate obtained using the following formula for the photoionization cross section of the hydrogen atom [79]:

$$\sigma = \frac{2^8 \pi e^2}{3c \sqrt{\varkappa_0}} \frac{1}{mM/(m+M)} \frac{1}{\mathcal{E}_i} \left(\frac{\mathcal{E}_i}{h\nu} \right)^{3/2}, \tag{41}$$

where we have to substitute the effective masses of an electron (m = $0.1m_0$) and a hole (M =

$0.3m_0$) in germanium, its permittivity ($\varkappa_0 = 16$), and the ionization energy of free excitons ($\mathscr{E}_{ex_i} = 4.0 \times 10^{-3}$ eV) deduced from the experimental spectra. An estimate of this kind obtained in the photon energy range $h\nu \approx \mathscr{E}_i$ gives a photoionization cross section $\sigma_{ex}^{theor} = 0.7 \times 10^{-13}$ cm^2. We can see that the photoionization cross section of indirect excitons in germanium deduced from the experimental data is in agreement with a theoretical estimate based on the simple hydrogen-like model. However, the shape of the short-wavelength edge of the exciton absorption spectrum differs from the dependence $\sigma \propto (h\nu)^{-8/3}$, which is predicted by the simple theory of the hydrogen-like atom. In the region of the short-wavelength edge, as in the case of the absorption by shallow impurities in germanium, the exciton absorption cross section decreases with rising photon energy much more slowly than predicted by this power law.

LITERATURE CITED

1.   Ya. I. Frenkel' (J. Frenkel), Phys. Rev., 37:17, 1276 (1931).
2.   G. H. Wannier, Phys. Rev., 52:191 (1937).
3.   S. A. Moskalenko, Fiz. Tverd. Tela, 4:276 (1962).
4.   J. M. Blatt, K. W. Böer, and W. Brandt, Phys. Rev., 126:1691 (1962).
5.   L. V. Keldysh and Yu. V. Kopaev, Fiz. Tverd. Tela, 6:2791 (1964).
6.   M. A. Lampert, Phys. Rev. Lett., 1:450 (1958).
7.   S. A. Moskalenko, Opt. Spektrosk., 5:147 (1958).
8.   L. V. Keldysh, Proc. Ninth Intern. Conf. on Physics of Semiconductors, Moscow, 1968, Vol. 2, publ. by Nauka, Leningrad (1968), p. 1303.
9.   L. V. Keldysh, Usp. Fiz. Nauk, 100:514 (1970).
10.   L. V. Keldysh, in: Excitons in Semiconductors [in Russian], Nauka, Moscow (1971), p. 5.
11.   R. S. Knox, Theory of Excitons, Suppl. 5 to Solid State Phys., Academic Press, New York (1963).
12.   J. H. Reuszer and P. Fisher, Phys. Rev., 135:A1125 (1964).
13.   R. L. Jones and P. Fisher, J. Phys. Chem. Solids, 26:1125 (1965).
14.   R. C. Milward and L. J. Neuringer, Phys. Rev. Lett., 15:664 (1965).
15.   A. I. Demeshina, R. L. Korchazhkina, N. N. Kuznetsova, and V. N. Murzin, Fiz. Tekh. Poluprovodn., 4:428 (1970).
16.   N. F. Mott, Trans. Faraday Soc., 34:500 (1938).
17.   G. Dresselhaus, J. Phys. Chem. Solids, 1:14 (1956).
18.   E. F. Gross and N. A. Karryev, Dokl. Akad. Nauk SSSR, 84:261, 471 (1952).
19.   W. Kohn, Solid State Phys., 5:275 (1957).
20.   T. P. McLean and R. Loudon, J. Phys. Chem. Solids, 13:1 (1960).
21.   D. Schechter, J. Phys. Chem. Solids, 23:237 (1962).
22.   W. Kohn and J. M. Luttinger, Phys. Rev., 97:883 (1955).
23.   S. Zwerdling, B. Lax, L. M. Roth, and K. J. Button, Phys. Rev., 114:80 (1959); S. Zwerdling, L. M. Roth, and B. Lax, J. Phys. Chem. Solids, 8:397 (1959); K. J. Button, L. M. Roth, W. H. Kleiner, S. Zwerdling, and B. Lax, Phys. Rev. Lett., 2:161 (1959).
24.   J. M. Luttinger and W. Kohn, Phys. Rev., 97:869 (1955); J. M. Luttinger, Phys. Rev., 102:1030 (1956).
25.   G. G. Macfarlane, T. P. McLean, J. E. Quarrington, and V. Roberts, Phys. Rev., 108:1377 (1957); Phys. Rev., 111:1245 (1958); J. Phys. Chem. Solids, 8:388 (1959).
26.   R. J. Elliott, Phys. Rev., 108:1384 (1957).
27.   K. L. Shaklee and R. E. Nahory, Phys. Rev. Lett., 24:942 (1970).
28.   E. F. Gross, V. I. Safarov, A. N. Titkov, and I. S. Shlimak, ZhETF Pis'ma Red., 13:332 (1971).

29. C. Benoit à la Guillaume and O. Parodi, Proc. Fifth Intern. Conf. on Physics of Semiconductors, Prague, 1960, publ. by Academic Press, New York (1961), p. 426.

30. J. R. Haynes, M. Lax, and W. F. Flood, J. Phys. Chem. Solids, 8:392 (1959).

31. L. V. Keldysh and A. N. Kozlov, Zh. Eksp. Teor. Fiz., 54:978 (1968).

32. E. A. Hylleraas, Phys. Rev., 71:491 (1947); E. A. Hylleraas and A. Ore, Phys. Rev., 71:493 (1947).

33. R. R. Sharma, Phys. Rev., 170:770 (1968); 171:36 (1968).

34. Ya. E. Pokrovskii and K. I. Svistunova, Fiz. Tekh. Poluprovodn., 4:491 (1970).

35. J. R. Haynes, Phys. Rev. Lett., 17:860 (1966).

36. V. S. Bagaev and L. I. Paduchikh, Fiz. Tverd. Tela, 13:484 (1971).

37. V. M. Asnin, A. A. Rogachev, and S. M. Ryvkin, Fiz. Tekh. Poluprovodn., 1:1742 (1967).

38. N. F. Mott, Philos. Mag., 6:287 (1961).

39. V. M. Asnin and A. A. Rogachev, ZhETF Pis'ma Red., 9:415 (1969).

40. Ya. E. Pokrovskii and K. I. Svistunova, ZhETF Pis'ma Red., 9:435 (1969).

41. Ya. E. Pokrovskii (Y. Pokrovsky), A. S. Kaminskii (A. Kaminsky), and K. I. Svistunova, Proc. Tenth Intern. Conf. on Physics of Semiconductors, Cambridge, Mass., 1970, publ. by US Atomic Energy Commission, Washington, DC (1970), p. 504

42. V. S. Bagaev, T. I. Galkina, O. V. Gogolin, and L. V. Keldysh, ZhETF Pis'ma Red., 10:309 (1969).

43. V. S. Bagaev, T. I. Galkina, and O. V. Gogolin, Kratk. Soobshch. Fiz., No. 2, 42 (1970); Proc. Tenth Intern. Conf. on Physics of Semiconductors, Cambridge, Mass., 1970, publ. by US Atomic Energy Commission, Washington, DC (1970), p. 500; Excitons in Semiconductors (ed. by B. M. Vul) [in Russian], Nauka, Moscow (1971), p. 19.

44. V. S. Vavilov, V. A. Zayats, and V. N. Murzin, ZhETF Pis'ma Red., 10:304 (1969).

45. C. Benoit à la Guillaume, F. Salvan, and M. Voos, J. Lumin., 1:315 (1970); Proc. Tenth. Intern. Conf. on Physics of Semiconductors, Cambridge, Mass., 1970, publ. by US Atomic Energy Commission, Washington, DC (1970), p. 516.

46. A. S. Kaminskii and Ya. E. Pokrovskii, ZhETF Pis'ma Red., 11:381 (1970).

47. J. D. Cuthbert, J. Lumin., 1:307 (1970).

48. B. M. Ashkinadze, I. P. Kretsu, S. M. Ryvkin, and I. D. Yaroshetskii, Zh. Eksp. Teor. Fiz., 58:507 (1970).

49. B. M. Ashkinadze, I. P. Kretsu, A. A. Patrin, and I. D. Yaroshetskii, Fiz. Tekh. Poluprovodn., 4:2206 (1970).

50. A. S. Alekseev, V. S. Bagaev, T. I. Galkina, O. V. Gogolin, N. A. Penin, A. N. Semenov, and V. B. Stopachinskii, ZhETF Pis'ma Red., 12:203 (1970).

51. A. S. Alekseev, V. S. Bagaev, T. I. Galkina, O. V. Gogolin, and N. A. Penin, Fiz. Tverd. Tela, 12:3516 (1970).

52. V. M. Asnin, A. A. Rogachev, and N. I. Sablina, ZhETF Pis'ma Red., 11:162 (1970).

53. C. Benoit a la Guillaume, M. Voos, F. Salvan, J. M. Laurent, and A. Bonnot, C. R. Acad. Sci. B, 272:236 (1971).

54. V. M. Asnin, A. A. Rogachev, and N. I. Sablina, Fiz. Tekh. Poluprovodn., 4:808 (1970).

55. Ya. E. Pokrovskii and K. I. Svistunova, ZhETF Pis'ma Red., 13:297 (1971).

56. V. N. Murzin and A. I. Demeshina, Opt. Spektrosk., 13:826 (1962).

57. A. I. Demeshina, V. A. Zayats, V. I. Lapshin, and V. N. Murzin, Zh. Prikl. Spektrosk., 13:346 (1970).

58. V. V. Buzdin, A. I. Demeshina, and V. N. Murzin, Prib. Tekh. Eksp., No. 1, 235 (1969).

59. B. V. Novikov, E. F. Gross, and M. A. Drygin, ZhETF Pis'ma Red., 8:15 (1968).

60. E. M. Voronkova, B. N. Grechushnikov, G. I. Distler, and I. P. Petrov, Optical Materials for Infrared Technology [in Russian], Nauka, Moscow (1965).

61. É. I. Zavaritskaya, Fiz. Tverd. Tela, 2:3009 (1960).

62. N. V. Zavaritskii and A. I. Shal'nikov, Prib. Tekh. Eksp., No. 1, 189 (1961).

63. A. A. Lipnik, Fiz. Tverd. Tela, 1:726 (1959); 3:2322 (1961).

64. V. S. Vavilov, V. A. Zayats, and V. N. Murzin, Proc. Tenth Intern. Conf. on Physics of
     Semiconductors, Cambridge, Mass., 1970, publ. by US Atomic Energy Commission,
     Washington, DC (1970), p. 509.
65. V. S. Vavilov, V. A. Zayats, and V. N. Murzin, in: Excitons in Semiconductors [in Russian],
     Nauka, Moscow (1971), p. 32.
66. Yu. L. Klimontovich and V. P. Silin, Usp. Fiz. Nauk, 70:247 (1960).
67. G. Mie, Ann. Phys. (Leipz.), 25:377 (1908).
68. H. C. van de Hulst, Light Scattering by Small Particles, New York, Wiley (1957).
69. K. S. Shifrin, Scattering of Light in Turbid Media [in Russian], Gostekhteoretizdat,
     Moscow—Leningrad (1951), p. 88.
70. O. V. Gogolin, Thesis for Candidate's Degree, Moscow State University (1970).
71. T. Holstein, Phys. Rev., 96:535 (1954).
72. W. G. Baber, Proc. R. Soc. A, 158:383 (1937).
73. R. N. Gurzhi and M. I. Kaganov, Zh. Eksp. Teor. Fiz., 49:941 (1965).
74. A. E. Stanevich and N. G. Yaroslavskii, Opt.-Mekh. Prom.-st', No. 5, 1 (1966).
75. T. S. Moss, Optical Properties of Semi-conductors, Butterworths, London (1959).
76. V. S. Vavilov, V. A. Zayats, and V. N. Murzin, Kratk. Soobshch. Fiz., No. 4, 9 (1971).
77. N. O. Lipari and A. Baldereschi, Phys. Rev. B, 3:2497 (1971).
78. M. Lax and E. Burstein, Phys. Rev., 100:592 (1955).
79. H. A. Bethe and E. E. Salpeter, Quantum Mechanics of One- and Two-Electron Atoms,
     Springer Verlag, Berlin (1958).

COLLECTIVE INTERACTIONS OF EXCITONS AND
NONEQUILIBRIUM CARRIERS IN
GALLIUM ARSENIDE AND SILICON *

L. I. Paduchikh

An investigation was made of the photoluminescence spectra of GaAs and Si in a wide range of optical excitation levels and at different temperatures. New information was obtained on the photoelectric properties of these semiconductors in the nonequilibrium carrier range in which the collective interaction of electrons and holes influenced strongly the energy band spectrum. The experimental results obtained were explained by postulating the appearance of a condensed phase consisting of constant-density electron–hole drops.

INTRODUCTION

The physical processes occurring in semiconductors in the presence of high nonequilibrium carrier densities are being studied intensively. Such studies are not only of fundamental importance but also of practical value. The operation of the majority of devices (for example, noncoherent light sources and lasers) is based on the use of semiconductors in which high densities of nonequilibrium electrons and holes are established.

At realistic nonequilibrium carrier densities ($N = 10^{17}$-10^{18} cm^{-3}) the Coulomb interaction between the carriers becomes comparable with their kinetic energy and it should govern the structure and electron properties of excited semiconductors in the same way as the conventional Coulomb interaction governs the structure of real solids and liquids. Thus, a system of nonequilibrium carriers can be regarded as a model of a real substance composed of electrons and ions, except that ions are now replaced with holes.

When the density of carriers in a semiconductor is low, electrons and holes may be bound into excitons. They differ basically from normal atoms by a low binding energy and a large Bohr radius. This makes it possible to vary their properties in a wide range of densities from the gaseous state to the state of strongly compressed matter and thus follow the transition from a dielectric exciton gas to a metallic Fermi gas of almost-free electrons and holes.

The range of intermediate intensities, when the distance between excitons is of the order of their Bohr radius, has hardly been investigated. According to the current theoretical ideas it is in this range of densities that we may expect to observe phenomena due to the collective interaction of excitons in semiconductors. A quantitative calculation of the energy spectrum of such a nonequilibrium system is difficult because in this range of densities there are funda-

---

* Thesis submitted for the degree of Candidate of Physicomathematical Sciences, defended in 1971 at the P. N. Lebedev Physics Institute, Academy of Sciences of the USSR, Moscow.

mental reasons why many parameters representing the band structure of semiconductors cannot be allowed for rigorously. In view of this, experimental investigations of nonequilibrium systems in this range have become particularly important.

The present paper reports an experimental investigation of the exciton spectra of semiconductors carried out in a wide range of nonequilibrium carrier densities.

CHAPTER I

COLLECTIVE INTERACTIONS OF EXCITONS IN SEMICONDUCTORS

At low nonequilibrium carrier densities the Coulomb interaction reduces the energy of free particles and produces a stable state characterized by the appearance of discrete levels in the forbidden band of a semiconductor. In the case of semiconductors with a high permittivity ($\varkappa \approx 10$) and low effective carrier masses ($m \approx 0.01m_0$–$01m_0$) we can expect formation of excitons with a binding energy $\mathcal{E}_0 \approx 10^{-2}$ eV (this is an order-of-magnitude estimate) and a large Bohr radius ($a \approx 10^{-6}$ cm). The values of $\mathcal{E}_0$ and a for each semiconductor can be found from the formulas $\mathcal{E}_0 = {}^1\!/_2 e^4 \mu / \varkappa^2 \hbar^2$ and $a = \varkappa \hbar^2 / \mu e^2$, where $\mu = m_e m_h / (m_e + m_h)$, m_e and m_h are the effective masses of an electron and a hole, and m_0 is the mass of a free electron.

As the exciton concentration n_{ex} rises, so that the distance between excitons becomes of the order of their Bohr radius, the kinetic energy of an electron or a hole $\hbar^2 n^{2/3} / m_{e,h}$ becomes comparable with the Coulomb energy of the electron–hole interaction $e^2 n^{1/3} / \varkappa$. When the nonequilibrium carrier density N reaches a value sufficient to satisfy the condition $Na^3 \geq 1$, excitons dissociate into free particles. The same conclusion of the dissocation of bound electrons and holes at high densities follows also from allowance for the screening of the Coulomb interaction. Thus, even at nonequilibrium carrier densities $N \approx 10^{17}$–10^{18} cm^{-3}, it is found that electrons and holes are strongly compressed so that a transition takes place from a dielectric gas of excitons to a metallic Fermi gas of almost-free electrons and holes.

The collective properties of excitons were first considered in connection with the possibility of their Bose condensation [1]. Since excitons have integral spin (each of them is composed of two Fermi particles), they may manifest properties typical of the Bose particles at low temperatures. According to [2], the critical temperature of the transition to the condensed state for an ideal Bose gas should be $kT_c = 1.8 \hbar^2 n_{ex}^{2/3} / M$, where $M = m_e + m_h$.

In view of the smallness of the effective mass of excitons, which is of the same order of magnitude as the electron mass, the Bose condensation should result in a sudden change of the temperature of the existence of a system of this kind, in contrast to the known and experimentally confirmed cases (such as that of superfluid liquid helium).

However, in view of the Fermi nature of electrons and holes, we find that at exciton concentrations such that the average distance between them becomes of the order of their Bohr radius, we cannot regard excitons as structure-free Bose particles and we must allow for the considerable influence of excitons on one another [3]. The final result of the transition to a new state depends on the nature of the interaction forces acting between excitons in the range of concentrations close to that given by $n_{ex} a^3 = 1$. If the repulsive forces predominate, we may expect the Bose condensation of excitons; if the attractive forces are stronger, we may expect [3] a transition of an exciton gas to a liquid phase composed of electron–hole drops. Such a transition should have all the features typical of a phase transition of the first kind and, in particular, there should be a critical temperature T_d below which the transition from one phase to the other is smooth and occurs without a discontinuity.

It is postulated in [3, 4] that the formation of electron–hole drops in a semiconductor does not exclude the intermediate stage of the appearance of exciton molecules. The existence of complexes composed of more than two particles is predicted in [5]. The hypothesis of the formation of biexcitons was put forward in [6]. Theoretical calculations [4-7] showed that even in the least favorable (from the point of view of formation of bound states) situation when the masses of an electron and a hole are identical, a molecule is still formed but it has a very small binding energy.

The hypothesis of the relative stability of biexcitons means that repulsive exchange forces between biexcitons predominate over the van der Waals attraction. However, as the concentration of biexcitons rises, the energy of each of them should increase, i.e., the binding energy should decrease. As shown in [4], at low biexciton concentrations this reduction is a linear function of the concentration in accordance with the formula $\mathcal{E}_b = \mathcal{E}_{b0} - \gamma \mathcal{E}_0 (Na^3)$, where $\mathcal{E}_{b0}$ is the binding energy of a biexciton in the $N \to 0$ limit and the coefficient in this expression is $\gamma > 10$.

We may thus have a situation in which, beginning from concentrations $Na^3 \approx 10^{-2}$, the existence of exciton molecules is less favorable from the energy point of view than the formation of liquid drops so that a phase transition similar to that discussed in [4] should be observed.

Theoretical estimates relating to these possibilities [3-7] are approximate also because the anisotropy of the effective masses of electrons and holes is ignored and this anisotropy may be one of the dominant parameters in each of the cases under consideration. Thus, we cannot rigorously predict the situation which should arise in semiconductors at high nonequilibrium carrier densities. In view of this one should try to obtain experimental information in order to select one of the hypotheses put forward.

The present author investigated two semiconductors: Si and GaAs. Semiconductors with indirect transitions and similar effective masses of electrons and holes are most promising from the point of view of experimental detection of the phase transition discussed here. Favorable conditions are provided by long lifetimes of indirect excitons, which are governed by the actual band structure. The combination of these characteristics in silicon, which is also characterized by a high binding energy of free excitons, led to the selection of this semiconductor.

The second semiconducting material was GaAs, which differs from Si by the coincidence of the valence and conduction band extrema in the k space and by the very different almost isotropic effective electron and hole masses. This semiconductor has found many practical applications, particularly in lasers. For this reason it would be interesting to study its optical properties at high nonequilibrium carrier densities.

In the experimental investigation of the collective interactions of excitons in Si and GaAs we found that until recently there has been no agreement even on such basic parameters as the forbidden band width E_g or the binding energy of free excitons $\mathcal{E}_0$. In the case of gallium arsenide this is explained by the difficultly of growing perfect single crystals with low impurity concentrations. In the case of silicon the exciton absorption lines in the region of indirect phonon-assisted transitions are not discrete and this makes it difficult to determine experimentally the binding energies of excitons from the absorption coefficient. For this reason part of the investigation was of auxiliary nature and it was carried out at relatively low nonequilibrium carrier densities, when electrons and holes were bound into excitons.

CHAPTER II

MEASUREMENT METHOD

Experimental investigations of excitons at high concentrations have been stimulated by the development of lasers, which are characterized by a great variety of output powers and emission wavelengths. Investigations of radiative recombination in semiconductors are currently one of the main methods for the determination of optical characteristics which can give considerable information on the energy-band structure and on the impurity and phonon spectra of crystals. The present paper describes a study of the physical processes which occur in semiconductors under strong optical excitation conditions. The use of strong optical excitation for the generation of nonequilibrium carriers has definite advantages over other known excitation methods. The photoluminescence spectra can be used to obtain most reliable information on the physics of recombination processes because optical excitation does not change significantly the crystal lattice structure.

§ 1. Optical System and Method of Recording Luminescence during Continuous Optical Excitation

A 40 mW Ne—He laser emitting at λ = 6328 Å (1.96 eV) was used as the excitation source.

We investigated the spectral composition of the luminescence of GaAs and Si in the range 1.4-1.52 and 1.05-1.1 eV, respectively. The photoluminescence of these semiconductors was recorded using two optical systems based on DFS-12 and MDR-2 diffraction-grating spectrometers designed for working in the photon ranges in question.

A block diagram of the apparatus is shown in Fig. 1. The measurements were carried out in the temperature range 2-100°K. Glass cryostats were used at liquid-helium and liquid-nitrogen temperatures. The temperature characteristics of the recombination radiation were determined using metal cryostats with optical windows. A sample was placed in a chamber through which liquid-helium vapor was pumped and the temperature of this vapor could be varied.

The first experiments showed that the apparatus should have the highest possible sensitivity in order to record weak photoluminescence emitted by undoped GaAs due to the gas laser excitation. An FÉU-28 photomultiplier was used as the detector of the GaAs luminescence.

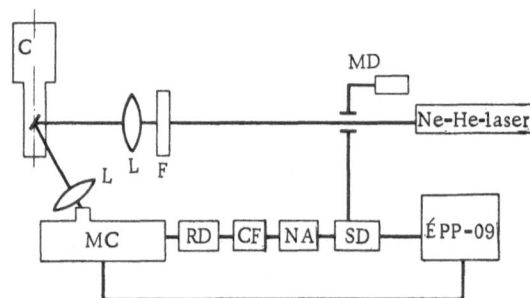

Fig. 1. Block diagram of the apparatus used to record the luminescence spectra of GaAs and Si during continuous optical excitation: C is a cryostat; MD is a modulator; L are focusing elements (lenses); F is a filter; MC is a monochromator; RD is a radiation detector; CF is a cathode follower; NA is a narrow-band amplifier; SD is a synchronous detector; ÉPP-09 is an automatic plotter.

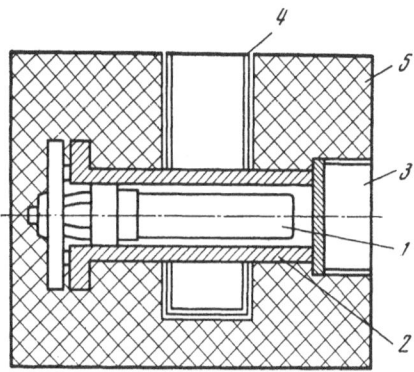

Fig. 2. System used to cool an FÉU-28 photomultiplier; 1) photomultiplier; 2) hermetic metal
jacket; 3) quartz cell; 4) can for liquid nitrogen;
5) thermal insulation.

The sensitivity threshold could be reduced by lowering the intrinsic noise of the FÉU-28 photomultiplier. The dark current of this photomultiplier was $i_T = 10^{-9}$ A at room temperature. According to [8], the amplitude of the photomultiplier noise can be written in the form $\overline{\Delta u^2} = 4kTR_e^*\Delta f + 2eiR_e^2\Delta f$, where R_e^* and R_e are the equivalent resistances representing the shot and current noise, and Δf is the pass band of the amplifier.

It was reported in [9] that at room temperature the noise in photomultipliers with oxygen-silver–cesium photocathodes was dominated by the current component but at low temperatures the minimum sensitivity was limited by the shot noise. When the cathode temperature was lowered from 300 to 77°K the dark current of the photomultiplier decreased by four or five orders of magnitude.

We designed a special cooling system for the FÉU-28 photomultiplier. The basic parts of this system are shown in Fig. 2. A metal jacket, which enclosed hermetically the photomultiplier, was cooled with nitrogen and the resultant convection and radiation emitted from the FÉU-28 photomultiplier reduced its temperature to $T \approx 80°K$.

The optimal load was selected bearing in mind that the noise due to the dark current of the photomultiplier should produce a voltage greater than that due to the thermal Johnson noise in the load. The current noise of the photomultiplier could be represented by an equivalent ohmic resistance of a metallic conductor R using the expression $2eiR^2\Delta f = 4kTR\Delta f$.

The equivalent noise resistance at T = 300°K was R = 51 MΩ. When the photomultiplier anode was connected to a load R > 300 MΩ, the minimum signal at the photomultiplier input was limited by the intrinsic noise.

The recombination radiation was recorded using the cooled FÉU-28 photomultiplier and the radiation was modulated at a low frequency f = 16 Hz. In order to match the FÉU-28 photomultiplier to the input of a narrow-band amplifier, we used a cathode follower based on a 6Zh1Zh vacuum tube in the zero-grid-current configuration [10, 11]. The anode of the FÉU-28 photomultiplier was supplied directly from KBGTs-0.35 batteries. The circuits and batteries were carefully screened in order to avoid the influence of stray alternating electric and magnetic fields.

The use of a cooled photomultiplier increased the sensitivity of the receiving and amplifying channel by almost three orders of magnitude at the frequency of the recombination radiation of GaAs.

A reverse-biased FD-111 germanium photodiode was used as a detector of recombination radiation emitted by Si. A dc source was shunted by a capacitor in order to reduce the grid and high-frequency strays. The signal picked up from the load was passed through a cathode follower and amplified with a U2-6 amplifier; it was then passed to a synchronous detector and plotted automatically.

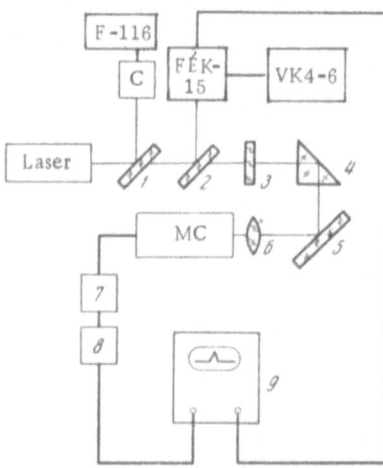

Fig. 3. Block diagram of the apparaturs used to record the luminescence spectra of GaAs and Si excited with short light pulses; C is a calorimenter; F-116 is a unit composed of an amplifier and a dc microvoltmeter; FÉK-15 is a photocell; VK4-6 is a digital voltmeter; 1, 2) rotatable plates; 3) neutral filter; 4) rotatable prism; 5) GaAs or Si semiconductor plate; 6) focusing lens; MC is a monochromator; 7) radiation detector; 8) wide-band amplifier; 9) oscillograph.

§2. Optical System and Method of Recording

Luminescence Due to High-Power Light Pulses

In order to investigate the photoluminescence of the two selected semiconductors in a wide range of excitation intensities, the author assembled a pulse apparatus which could produce photon flux densities up to $I \approx 10^{24}$ cm$^{-2} \cdot$sec^{-1}. In this case the excitation source was a ruby laser. Figure 3 shows the block diagram of the apparatus. A ruby laser beam was directed onto the investigated semiconductor and the luminescence emitted by the semiconductor passed through a dispersing element coupled to a suitable detector. The signal produced by the load of the recording system was amplified with a wide-band amplifier, displayed on the screen of an oscillograph, and then recorded photographically on a high-sensitivity film.

Under pulse excitation conditions the free oscillation in the ruby laser lasted $\tau_p = 200$ μsec. Shorter pulses were obtained by Q switching. The switch was a glass prism rotating at a rate of 10,000 rpm. Single-spike emission was obtained when a KS-19 filter was placed between this prism and the ruby crystal. In this case the pulse duration was 20 nsec.

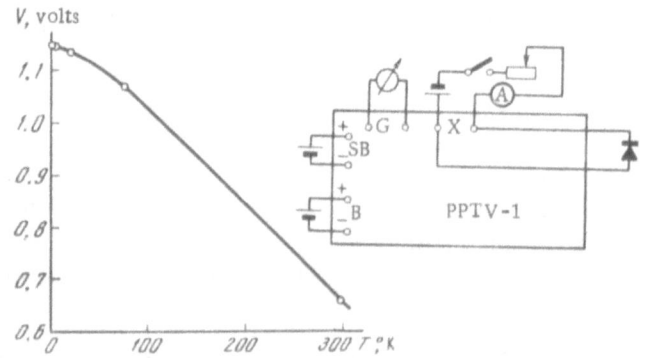

Fig. 4. Calibration curve of a thermometer based on a diffused GaAs diode. The circuit used in temperature measurements is shown on the right: PPTV-1 is a high-resistance dc potentiometer; X are terminals for voltage measurements; G are galvanometer terminals; SB are standard terminals; B are main-battery elements; A is a microammeter.

The laser radiation intensity was monitoried either with a calorimeter, whose signal was amplified and recorded with an F-116 unit, or with an FÉK-15 photocell connected to a VK4-6 digital voltmeter.

The laser radiation intensity was reduced using calibrated neutral filters.

§3. Temperature Measurement Method

The photoluminescence spectra of Si and GaAs were investigated in the temperature range 2-100°K. Cooling from 4.2 to 2°K was achieved by reducing the saturated helium vapor pressure. The temperature was measured by comparing the readings of a manometer with the known boiling temperature of liquid helium at the appropriate vapor pressure.

Above helium temperatures, we used thermometers based on diffused diodes. The voltage drop across a diode made of n-type GaAs was an almost linear function of temperature between 4.2 and 300°K [12, 13]. The voltage was measured using a conventional compensation circuit (Fig. 4). Figure 4 gives the calibration curve of the GaAs diode. In this way the temperature was measured to within 0.1 deg K. The characteristics of these diodes were highly reproducible.

§4. Determination of Temperature Rise in a
Semiconductor during Continuous Optical Excitation

When a semiconductor is excited with photons of energy exceeding its forbidden-band width, some of the optical energy is converted into heat. The temperature rise in the active region of a semiconductor, where the recombination processes take place, generally depends not only on the thermal power dissipated in the path of the beam, geometrical dimensions of the sample, and the cooling method, but also on the temperature of the medium in which the semiconductor is located. This is due to the change in the temperature gradient between the heated region and the cooling agent and due to the considerable temperature dependence of the thermal conductivity of semiconductors k (for example, $k_{GaAs} \approx 1$ W·cm^{-1}·deg^{-1}, $k_{Si} \approx 0.2$ W·cm^{-1}·deg^{-1} at T = 2°K and $k_{GaAs} \approx 15$ W·cm^{-1}·deg^{-1}, $k_{Si} = 60$ W·cm^{-1}·deg^{-1} at T = 20°K [14]). For this reason it may happen that whereas at low temperatures the removal of heat from the active region of a semiconductor is due to heat exchange with a cooling agent near the illuminated surface, at high temperatures the rise of the temperature of the sample is governed mainly by the removal of heat from the bulk of the sample. This happens if the thermal contact between the semiconductor and the cooling agent is lost. The maximum power which can be dissipated by a heated surface without the loss of the thermal contact is 0.5 W/cm^2 in liquid helium [15] and 10 W/cm^2 in liquid nitrogen [16].

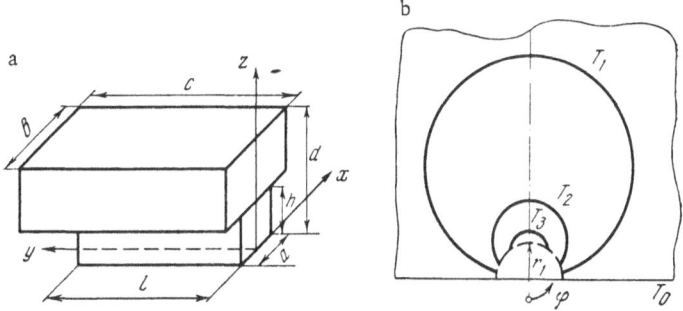

Fig. 5. Model used in calculating temperature rise: a) uniformly illuminated surface of a semiconductor plate; b) active region of a semiconductor illuminated with a strongly focused cw laser beam.

In order to calculate the temperature rise of the surface of an illuminated semiconductor in the case when the heat is removed mainly from the bulk of the crystal, we can use the same approach as that employed in calculating the rise of the temperature of the p–n junction plane in a continuously operating semiconductor laser [17]. The model used in such calculations is shown in Fig. 5a. It is assumed that the illuminated surface of a crystal (la) receives a uniform heat flux Q. Then, the temperature distribution in the semiconductor can be found by solving the following homogeneous heat conduction equation:

$$T_{xx} + T_{yy} + T_{zz} = 0 \tag{1}$$

which is subject to the boundary conditions on the semiconductor plate,

$$T^1\left(\begin{matrix} -\frac{a}{2} \\ \frac{a}{2} \end{matrix}, y, z\right) = T_0,$$

$$T^1\left(x, \begin{matrix} 0 \\ l \end{matrix}, z\right) = T_0,$$

$$T^1_z(x, y, 0) = -\frac{Q}{k_1}$$

and on the metal heat sink,

$$T^2\left(\begin{matrix} -\frac{b}{2} \\ \frac{b}{2} \end{matrix}, y, z\right) = T_0,$$

$$T^2(x, y, d) = T_0,$$

$$T^2\left(x, \begin{matrix} 0 \\ c \end{matrix}, z\right) = T_0.$$

where a, l, h are the dimensions of the semiconductor and b, c, (d − h) are the dimensions of the heat sink (Fig. 5); T_0 is the temperature of the ambient medium.

The matching conditions at the metal–conductor contact are of the form

$$T^2(x, y, h) = \begin{cases} T_0, & \left|\frac{a}{2}\right| \leqslant x \leqslant \left|\frac{b}{2}\right|, \ 0 \leqslant y \leqslant c, \\ T^1(x, y, h), & -\frac{a}{2} \leqslant x \leqslant \frac{a}{2}, \quad 0 \leqslant y \leqslant l, \\ T_0, & -\frac{b}{2} \leqslant x \leqslant \frac{b}{2}, \quad l \leqslant y \leqslant c, \end{cases}$$

$$k_1 T^1_z(x, y, h) = k_2 T^2_z(x, y, h), \quad -\frac{a}{2} \leqslant x \leqslant \frac{a}{2} \ \ 0 \leqslant y \leqslant l,$$

where k_1 is the thermal conductivity of the semiconductor and k_2 is the thermal conductivity of the metal heat sink.

The solution of the above equation is a series with an alternating sign:

$$T^1(x, y, z) = T_0 + \Delta T(x, y, z) = T_0 + \sum_{n=0}\sum_{k=1} \frac{8Q}{k_1\sqrt{\lambda_n^2+\mu_k^2}} \frac{(-1)^n[1-(-1)^k]}{\pi^2 k(2n+1)} \times$$

$$\times \left\{ \frac{e^{-\sqrt{\lambda_n^2+\mu_k^2}h}\left[\frac{k_1}{k_2}\tanh\sqrt{\lambda_n^2+\mu_k^2}\,(d-h) - \frac{al}{bc}\right]}{\cosh\sqrt{\lambda_n^2+\mu_k^2}h\left[\frac{k_1}{k_2}\tanh\sqrt{\lambda_n^2+\mu_k^2}(d-h)\tanh\sqrt{\lambda_n^2+\mu_k^2}h + \frac{al}{bc}\right]}\cosh\sqrt{\lambda_n^2+\mu_k^2}z + e^{-\sqrt{\lambda_n^2+\mu_k^2}z}\right\}\cos\frac{\pi(2n+1)}{a}x\sin\frac{\pi k}{l}y,$$

$$\tag{2}$$

where

$$\lambda_n = \frac{\pi(2n+1)}{a}, \qquad n = 0, 1, 2 \ldots,$$

$$\mu_k = \frac{\pi k}{l}, \qquad\qquad k = 1, 2, 3 \ldots$$

The series (2) converges rapidly in respect of n and k and in approximate calculations it is sufficient to retain only the first term. The higher terms give rise to a correction which is smaller than ten percent of the first term in the sum. The solution of the equation obtained in the first approximation of the series expansion can be written in the form

$$\Delta T(x, y, z) = \frac{16Qla}{\pi^5 k_1 \sqrt{a^2 + l^2}} \times \left\{ \frac{e^{-\frac{\pi h}{al}\sqrt{a^2+l^2}} \left[\frac{k_1}{k_2} \tanh \frac{\pi(d-h)}{al}\sqrt{a^2+l^2} - \frac{al}{bc} \right]}{\cosh \frac{\pi h}{al}\sqrt{a^2+l^2}\left[\frac{k_1}{k_2} \tanh \frac{\pi(d-h)}{al}\sqrt{a^2+l^2} \tanh \frac{\pi d}{al}\sqrt{a^2+l^2} + \frac{al}{bc} \right]} \times \right.$$

$$\left. \times \left| \cosh \frac{\pi}{al}\sqrt{a^2+l^2}z + \exp\left(-\frac{\pi\sqrt{a^2+l^2}}{al}z\right) \right\} \cos \frac{\pi}{a}x \sin \frac{\pi}{l}y. \tag{3}$$

In this approximation the strongest temperature rise occurs on the illuminated surface of the semiconductor and it is this rise ΔT that should be calculated. The value of ΔT can be found from Eq.(2) by substituting z = 0. The average temperature rise of the illuminated surface can thus be found from the formula

$$\overline{\Delta T}(x, y, 0) = \frac{1}{al}\int\limits_{-\frac{a}{2}}^{\frac{a}{2}} dx \int\limits_{0}^{l} \Delta T(x, y, 0)\, dy. \tag{4}$$

The determination of the temperature of the illuminated surface of a semiconductor in a situation when heat exchange with the ambient medium occurs only through the side faces of the semiconductor plate is a special case of the general solution (2) and it can be represented by

$$\overline{\Delta T}(z = 0) = \frac{64Q}{k_1 \pi^5} \frac{al}{\sqrt{a^2+l^2}} \left[\frac{e^{-\frac{\pi h}{al}\sqrt{a^2+l^2}}}{\sinh \frac{\pi h}{al}\sqrt{a^2+l^2}} + 1 \right]. \tag{5}$$

A considerable proportion of the measurements of the luminescence spectra of GaAs and Si during continuous excitation with the Ne—He laser radiation was carried out with the laser beam strongly focused so that the size of the illuminated spot was comparable with the diffusion length of nonequilibrium carriers in the semiconductor. These measurements were usually carried out with the semiconductor immersed in liquid helium whose temperatures was 2-4.2°K when the thermal conductivity of the semiconductor was relatively low. In this case all the heat exchange processes occurred near the surface of the semiconductor in a layer of thickness equal to the depth of optical absorption of the Ne—He laser radiation or at distances equal to the diffusion length of nonequilibrium carriers L_d, when the quantum efficiency of the recombination radiation was $\eta \ll 1$.

In this situation the temperature rise in a semiconductor can be estimated by solving the steady-state heat conduction equation of the type

$$\frac{1}{r}\frac{\partial}{\partial r}\left(r\frac{\partial T}{\partial r}\right) + \frac{1}{r^2}\frac{\partial^2 T}{\partial \varphi^2} = 0 \tag{6}$$

subject to the following boundary conditions (Fig. 5b):

$$\frac{\partial T}{\partial r}(r_1, \ \varphi) = -\frac{Q}{k}; \qquad T\left(r, \ \begin{array}{c}\frac{\pi}{2}\\ \frac{3\pi}{2}\end{array}\right) = T_0; \qquad T(\infty, \ \varphi) = T_0.$$

Equation (6) can be solved by assuming that the propagation of heat in the half-space $\pi/2 \leq \varphi \leq 3\pi/2$ is spherically symmetric and, secondly, that all the heat is evolved on the surface of a semiconductor in an area of r_1, where $2r_1$ is the diameter of the light spot.

The solution of this equation can be represented, like the solution (2), in the form of a series

$$T(r, \ \varphi) = T_0 + \Delta T(r, \ \varphi) = \frac{2Q}{\pi k} \sum_{n=1}^{\infty} \frac{r_1^{n+1}}{n^2 r^n} (-1)^{2n+1} \cos n\varphi. \tag{7}$$

The series converges rapidly with respect to n (proportionally to $1/n^2$) so that the solution of Eq. (6) can be written in the form

$$\Delta T = \frac{2Q}{k\pi} \frac{r_1^2}{r} |\cos \varphi|, \qquad \frac{\pi}{2} \leqslant \varphi \leqslant \frac{3\pi}{2}. \tag{8}$$

In the case of a uniform distribution of heat on the surface of a hemisphere of radius r_1 the constant temperature curves (isotherms) in the $(r, \ \varphi)$ plane are of the type shown in Fig. 5b. The hottest spot in this case is the point whose coordinates are $(r_1, \ \pi)$ and it is interesting to estimate the temperature T at this point. The estimated temperature should be close to the real average temperature in the case of continuous absorption of radiation in a sample. Therefore, we shall assume that the density of heat sources in a hemisphere of radius r_1 is less than the corresponding density on an illuminated surface of a crystal by a factor equal to the ratio of the areas of the two surfaces. Then, the formula used to find ΔT can be written in the form

$$\overline{\Delta T} = \frac{Q_S}{k\pi r_1}, \tag{9}$$

where Q_S is the total heat evolved on the illuminated surface.

§5. Determination of Temperature Rise in a Semiconductor during Illumination with High-Power Light Pulses

In estimating the temperature rise in crystals illuminated with light pulses we must consider such parameters as the duration of a light pulse τ_p, nonequilibrium carrier lifetime τ, optical absorption depth L, diffusion length of nonequilibrium carriers L_d, and thermal diffusion length s.

The concept of the thermal diffusion length is introduced in [18]. It represents a distance travelled by a thermal front in a time τ_p. The thermal diffusion length can be calculated from the formula [18]

$$s = \sqrt{\frac{\pi k \tau_p}{\rho c}}, \tag{10}$$

where k, ρ, and c are the thermal conductivity, density, and specific heat of the semiconductor, respectively.

The substitution of the numerical values of k, ρ, and c into Eq. (10) gives dependences of the type s_{GaAs} [cm] = $3.3\,\tau_p^{1/2}$ [sec] at T = 80°K [18] and s_{GaAs} [cm] = $150\,\tau_p^{1/2}$ [sec] at T = 2°K.

Below 4.2°K the thermal diffusion length in silicon may be estimated from s_{Si} [cm] = 120 $\tau_p^{1/2}$ [sec].

In the optical excitation of semiconductors with a ruby laser the duration of a thermal perturbation was equal to the pulse duration τ_p = 200 μsec. This duration τ_p was known to exceed all the time constants of semiconductors at low temperatures. At T = 2°K the thermal diffusion length was s = 2.1 cm in GaAs, which was considerably greater than L or L_d. For this reason the results of calculations obtained for continuous illumination could be used to estimate the temperature rise due to light pulses.

The situation was different when the active region of a semiconductor was illuminated with high-power short pulses of the type emitted by Q-switched solid-state lasers. In this situation, τ_p was comparable with the low-temperature lifetime τ in GaAs and in the case of Si the pulse duration was considerably shorter than the lifetime of nonequilibrium electron–hole pairs. Moreover, in the case of silicon the value of s was found to be comparable with L and L_d. In view of this a rough estimate of the temperature rise in the active region of a semiconductor could be obtained in the adiabatic approximation from the formula

$$Q_\tau \text{ [J]} = \int_{T_0}^{T_1} c\,(T)\,\bar{m}\Delta T, \tag{11}$$

where $\bar{m}$ is the mass of the semiconductor calculated from the volume governed by L_d and by the area of the illuminated surface, and Q_τ is the thermal energy received by the semiconductor in a time τ_p.

The results of calculation of $\overline{\Delta T}$ carried out using Eqs. (5), (9), and (11) are given in Table 1 for some specific situations encountered in our experiments.

In the calculation of $\overline{\Delta T}$ it was assumed that the radiative quantum efficiency was $\eta \ll 1$ and, therefore, the maximum temperature rise of the semiconductor expected in the experiments was calculated. Since the change in the thermal conductivity of a semiconductor during heating was ignored, the values of ΔT obtained for the majority of high-power optical excitation cases could be overestimated. Nevertheless, the calculated results (Table 1) provided a quantitative confirmation that high optical energy densities in semiconductors should be produced either by strong focusing of light beams (in the continuous excitation case) or by the use

TABLE 1

T, °K	Semi-conductor	k, W·cm⁻¹·deg⁻¹	Laser	Laser operation	Q, W/cm²	Formula	ΔT, °K	Other parameters
2	GaAs	1	Ne — He	Continuous	0.5	(5)	0.05	$a = 0.5,\ l = 0.7,\ h = 0.2$ cm
	GaAs	1	Ne — He	Continuous	$1.3\cdot10^4$	(9)	13	$Qs = 4\cdot10^{-2}$ W, $r_1 = 10^{-3}$ cm
	Si	0.2	Ne — He	Free oscillation	$1.3\cdot10^2$	(9)	6	$Qs = 4\cdot10^{-2}$ W, $r_1 = 10^{-2}$ cm
4.2	GaAs	10	Ruby	Free oscillation	10^3	(5)	40	$a = 0.5,\ l = 0.5,\ h = 0.2$ cm
	Si	—	Ruby	Q-switched	10^6	(11)	10	$p = 2.33$ g/cm³ $C = aT^3$, where $a = 1.2\cdot10^{-7}$ cal·g⁻¹·deg⁻⁴, $L_d = 2\cdot10^{-2}$ cm

of short light pulses. The latter have definite advantages if it is desirable to reduce $\overline{\Delta T}$, particularly in the case of those semiconductors which are characterized by short nonequilibrium carrier lifetimes $\tau\,(\tau \ll \tau_p)$.

CHAPTER III

PHOTOLUMINESCENCE OF GALLIUM ARSENIDE

§1. Excitons in GaAs and Their Role in Radiative Recombination

Gallium arsenide is a III-V semiconducting compound with spherically symmetric energy bands. The absolute minimum of the conduction band is located at the center of the Brillouin zone. The valence band has a typical semiconductor structure, consisting of three subbands characterized by different effective hole masses.

It is shown in [19] that, at low temperatures in direct-gap semiconductors, there is a high probability of formation of free excitons followed by their radiative recombination until kT becomes equal to the binding energy. In the simple hydrogen-like model the binding energy of unexcited state of an exciton in GaAs is $\mathscr{E}_0 = 4.4$ meV and the radius of the first Bohr orbit is $a = 110$ Å.

Free excitons in GaAs were first discovered in [20] by a study of the optical absorption spectra at low temperatures. It was shown in [21] that the Coulomb interaction between electrons and holes not only produces discrete states near an allowed energy band but also alters the absorption coefficient in the range of energies exceeding the forbidden-band width. The experimental values of the binding energy of excitons and of the forbidden-band width were found in [20] using the theory of the optical absorption in the case of two nondegenerate spherical bands, for which a well-known analytic solution is available [21]. The optical absorption curves obtained below 90°K yielded $\mathscr{E}_0 = 3.4$ meV and $a = 190$ Å.

At the time when the present study was begun, there were only two published papers on the edge luminescence spectra of undoped GaAs crystals [22, 23]. A weak photoluminescence of energy close to 1.514 eV was observed below 20°C and a stronger luminescence was found at 1.493 eV. The former was attributed to the recombination of free excitons in GaAs and the latter to the radiative capture of electrons from the conduction band by acceptor centers in GaAs. The half-width of the short-wavelength luminescence exceeded considerably the expected value because of the thermal smearing of the energy of recombining particles, and the energy of this luminescence differed from the forbidden band width by an amount exceeding the binding energy of free excitons.

In crystals with many many imperfections any other elastic or inelastic interaction nonequilibrium carriers with lattice virbations or with impurities could broaden the absorption and luminescence lines. In the case of direct-gap semiconductors such interactions would give rise to new selection rules for quasimomentum so that transitions with $k \neq 0$ should become possible.

The high concentrations of impurities in the samples used in [20, 22, 23] could not only broaden considerably the luminescence or absorption lines and hence give rise to a considerable scatter of the absolute values of the binding energy of excitons and the forbidden-band width, but also could mask other processes with narrower luminescence lines.

The possibility of the existence of exciton-impurity complexes was first considered in [5]. A general solution of the wave equation with a potential energy allowing simultaneously for the interaction of an electron and a hole with an impurity was found to be practically unattainable. The solution could be obtained only for certain limiting cases of the ratio of the effective masses or in the case when the characteristics of the band structure were ignored.

The dissociation energy of a complex was defined as the minimum energy necessary to detach an exciton from an impurity. Calculations of the energy of large-radius complexes, i.e., of excitons bound to shallow singly charged impurities, were reported in [24-26]. It was shown in [25] that the binding energy of an exciton to an ionized donor $\mathcal{E}_{XD^+}$ in a semiconductor characterized by an effective mass ratio $m_e/m_h = 0.15$ was $\mathcal{E}_{XD^+} = 1.06\,\mathcal{E}_D$, where $\mathcal{E}_D$ is the ionization energy of a neutral donor. In the case of an exciton captured by an ionized acceptor it was found that $\mathcal{E}_{XA^-} = 1.4\,\mathcal{E}_A$. The energy needed to detach an exciton from a neutral donor was computed to be $\mathcal{E}_{XD^0} = 0.13\,\mathcal{E}_D$. It was shown in [26] that $\mathcal{E}_{XD^0} = 0.23\,\mathcal{E}_D$, and $\mathcal{E}_{XA^0} = 0.07\,\mathcal{E}_A$. According to the estimates given in [25], the luminescence energies of bound excitons were of the form

$$\hbar\omega\,(XD^0) = E_g - \mathcal{E}_0 - 0.13\mathcal{E}_D,$$
$$\hbar\omega\,(XD^+) = E_g - \mathcal{E}_D - 0.06\mathcal{E}_D,$$
$$\hbar\omega\,(XA^0) = E_g - \mathcal{E}_0 - 0.07\mathcal{E}_A,$$
$$\hbar\omega\,(XA^-) = E_g - \mathcal{E}_A - 0.4\mathcal{E}_A.$$

The effective-mass approximation was applied to the ground and first excited states of shallow acceptors and donors in GaAs and estimates of the luminescence energies of bound excitons were obtained in [27]: these energies were 1.5147, 1.5133, 1.491, and 1.481 eV, in the order given above. Thus, at low temperatures the luminescence spectrum of GaAs should be a combination of lines and luminescence bands due to different recombination processes.

It is clear that the existence of impurity complexes should depend very critically on the optical excitation rate and on the temperature of a semiconductor because the dissociation energy of such complexes is low and the effective Bohr radius of excitons is large.

The first reports of the experimental observation of narrow lines in the luminescence spectrum of GaAs were published in [27, 28] and they will be compared later with the results of our investigation of the luminescence spectra.

The present investigation was carried out in order to obtain information on collective interactions of excitons in GaAs. However, in the course of this study we found that it was quite difficult to detect the presence of free excitons on the basis of the low-temperature photoluminescence of GaAs because the intensity of this luminescence was very low.

We studied the radiative recombination processes in GaAs at different rates of optical excitation and over a wide temperature range. A considerable proportion of the measurements was carried out at high rates of carrier generation when the interaction between electrons and holes was the dominant factor in the recombination process.

§2. Investigation of Luminescence Spectra of GaAs

at Different Optical Excitation Rates and Helium Temperatures

We investigated the photoluminescence of undoped gallium arsenide crystals in the range of energies close to the forbidden band width. Samples of GaAs had n-type conduction and differed in the amounts of accidental impurities which were captured during growth. Since we used crystals grown by different methods, including those prepared by liquid epitaxy, the samples were selected in accordance with the edge luminescence spectra T = 4.2°K. The carrier density and mobility at room and liquid-nitrogen temperatures were determined in the laboratories where these crystals were grown. We only carried out measurements of the Hall

TABLE 2

GaAs batch No.	T = 300°K		T = 77°K		Preparation method
	μ, cm$^2 \cdot$ V$^{-1} \cdot$ sec^{-1}	n, cm^{-3}	μ, cm$^2 \cdot$ V$^{-1} \cdot$ sec^{-1}	n, cm^{-3}	
1	6100	$1.6 \cdot 10^{16}$	10700	$6.1 \cdot 10^{15}$	Bridgman
2	5300	$2.3 \cdot 10^{14}$	20000	$1.1 \cdot 10^{14}$	Liquid epitaxy
3	5900	$1.5 \cdot 10^{15}$	25000	10^{15}	Liquid epitaxy
4	6700	$6 \cdot 10^{15}$	27000	$5.3 \cdot 10^{15}$	Liquid epitaxy

effect in crystals grown from the melt. We obtained experimental results showing a certain regularity in the photoluminescence spectra, which enabled us to interpret some of the radiative transitions in GaAs at different temperatures and photoexcitation rates.

Table 2 gives the values of the majority-carrier mobility and density in the investigated samples.

Since the absorption coefficient of GaAs at wavelengths corresponding to the exciting radiation band was about 3×10^4 cm^{-1}, the absorption was concentrated in the surface layer. The surface defects generated during mechanical grinding and polishing contributed considerably to the known radiative recombination of nonequilibrium carriers. For this reason the surface layer was removed in a polishing etchant suitable for n-type materials, whose composition was $5H_2SO_4 : 2H_2O : 1H_2O_2$. The rate of etching at room temperature was 6 μ/h.

In the derivation of the principal formulas in the Wannier model it is assumed that the valence and conduction bands of a semiconductor are parabolic. In view of the principle of indeterminacy, this approximation is not very appropriate for large-radius excitons because only a small part of the k space is needed in the derivation of the wave function. Therefore, in the case of crystals with a high permittivity and coincident absolute extrema of the allowed bands in the k space, the Wannier model and the hydrogen-like spectrum are confirmed satisfactorily by the experimental results. The parameters of GaAs make it an almost ideal semiconductor satisfying these requirements and, therefore, the estimates of the binding energy of free excitons and of its Bohr radius obtained in the Wannier approximation should be confirmed by the experimental results. According to the Wannier model, the luminescence due to ground-state excitons in GaAs should be located at E = 1.5156 eV. However, the first studies of the photoluminescence carried out at low temperatures indicated that the GaAs luminescence spectra could not be explained in this simple manner.

Figure 6 (curve 1) shows the luminescence spectrum of GaAs crystals (Table 2) obtained at T = 4.2°K using a photon flux I $\approx 10^{18}$ cm$^{-2} \cdot$ sec^{-1}. Two main luminescence lines with energies $E_1 = 1.514 \pm 0.0004$ eV (λ = 8188 Å) and $E_2 = 1.4927 \pm 0.0004$ eV (λ = 8310 Å), respectively,

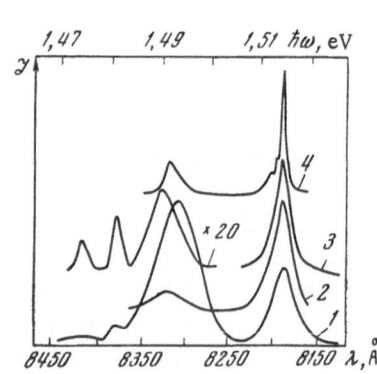

Fig. 6. Photoluminescence spectra of GaAs at T 4.2°K. The numbers of the curves represent the batches of GaAs samples whose properties are listed in Table 2; the abscissa of each curve is displaced in an arbitrary manner.

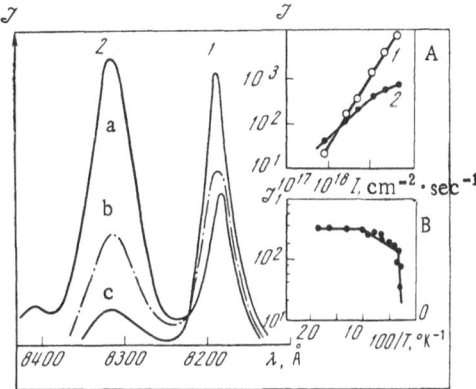

Fig. 7. Photoluminescence spectra of a GaAs sample No. 1 obtained at different optical excitation rates (cm^{-2}·sec^{-1}): a) 8×10^{17}; b) 2.2×10^{18}; c) 6×10^{19}. An analysis of these spectra yields the results shown on the right: A) dependence of the photoluminescence intensity of GaAs on optical excitation rate at 4.2°K ($1 - 8188$ Å; $2 - 8310$ Å); B) temperature dependence of the luminescence intensity $\mathscr{J}_1$ for $I = 5 \times 10^{17}$ cm^{-2}·sec^{-1}.

were observed. The two lines had almost symmetric profiles and half-widths considerably greater than the spectral slit width or the broadening caused by the thermal smearing of the luminescence spectra. The half-width of the short-wavelength photoluminescence line was 0.004 eV (≈ 10 kT) and the corresponding half-width of the long-wavelength line was twice as large (0.0076 eV). No significant redistribution of the intensities of the two lines or changes in the absolute intensities of the two lines were observed in the temperature range 2–4.2°K. Similar spectra were recorded for GaAs grown from the melt.

The distribution of the luminescence intensities of the two lines depended strongly on the optical excitation rate (Fig. 7) [29]. The nature of the dependence of the intensity $\mathscr{J}_{E_2}$ of the line E_2 on the rate of excitation I was typical of the recombination of nonequilibrium carriers assisted by the presence of impurity centers (Fig. 7a). Since the long-wavelength luminescence disappeared at T > 30°K [28] and the energy of this luminescence was less than the forbidden-band width by 27 meV, we expected the 1.493 eV luminescence to be due to the impurity recombination involving the participation of shallow donors and acceptors. It was reported in [27, 30] that specially doped GaAs crystals grown by liquid epitaxy exhibited a splitting of this luminescence into several components and it was found that the main contribution to the luminescence was due to the recombination of carriers associated with different pairs of donors and acceptors.

When the photoexcitation rate was increased, the luminescence intensity at 1.514 eV varied as $I^{1.5}$ and in the range $I > 2 \times 10^{18}$ cm^{-2}·sec^{-1} this line dominated the spectrum (Fig. 7A). The fall of the intensity $\mathscr{J}_1$ in the temperature range 10–20°K obeyed the law $\mathscr{J}_0(1 - e^{-\zeta_1/kT})$, where $\zeta_1 = 0.82$ meV and $\mathscr{J}_0$ is the luminescence intensity at T = 2–4.2°K. Above T = 25°K the dependence of $\mathscr{J}_1$ on T was stronger: the intensity was proportional to $\exp(\zeta_2)$, where $\zeta_2 \approx 9.5$ meV. At optical excitation rates $I > 10^{21}$ cm^{-2}·sec^{-1} the short-wavelength line was observed right up to temperatures close to room temperature.

Since large-radius excitons are macroscopic particles capable of moving across a crystal, their interaction with phonons and impurities may dominate the distribution of the kinetic energies of excitons in the exciton band. According to [31], in the case of direct-gap semiconductors, the short-wavelength edge of the luminescence line of the Wannier excitons has the Lorentzian profile with a half-width $\Delta = h\Gamma(\mathbf{k})_{k=0}$, where $\Gamma(\mathbf{k})$ is a quantity which is the reciprocal of the relaxation time representing the scattering of excitons by lattice vibrations.

This is true in the range of moderately low temperatures and for predominant interaction of free excitons with acoustic phonons. The sudden drop in the long-wavelength edge occurs because of the absence of allowed states below the exciton level with $\mathbf{k} = 0$.

Theoretically, the intensity in the long-wavelength edge of an absorption band falls to zero at a distance from the peak equal to about one-quarter of its half-width. It is shown in [32] that excitons are scattered similarly to electrons and holes but the interaction is much weaker.

The disagreement between the energy of the short-wavelength luminescence maximum calculated in accordance with the Wannier model and the experimental results as well as the enormous spectral half-width Δ are in conflict with the current theoretical ideas on the radiative recombination of free excitons in direct-gap semiconductors [19, 31].

Figure 6 shows a series of photoluminescence curves of GaAs crystals with different impurity concentrations which decrease from sample 1 to sample 4. The absolute radiative efficiency in the region of the short-wavelength line increases and its half-width decreases as the impurity concentration is reduced.

In the case of crystals characterized by a high mobility of free carriers at liquid nitrogen temperature the luminescence spectrum has a definite structure (curve 4 in Fig. 6). For $I \approx 10^{18}$ cm^{-2}·sec^{-1} at T = 2°K the luminescence $\mathscr{J}_1$ with a spectral width $\Delta = 5.5$ Å predominates. At lower rates of generation of nonequilibrium photocarriers ($I \approx 7 \times 10^{15}$ cm^{-2}·sec^{-1}) only one narrow luminescence line remains in the spectrum and the half-width Δ of this line is limited by the spectral slit width (Fig. 8).

A considerable increase in the intensity of the optical excitation causes a strong rise of the long-wavelength wing and in the background of this wing the following recombination maxima appear in succession: 1.5137 eV (8191.7 Å); 1.5134 eV (8193.4 Å); 1.5133 eV (8194 Å); 1.5125 eV (8201 Å).

The spectral width of the DFS-12 spectrometer slit used to record spectrum 4 in Fig. 8 was 0.05 meV.

The appearance of strong narrow lines in the photoluminescence spectrum of GaAs provided an experimental confirmation of the theoretical prediction of anomalously large oscillator strengths for the transitions due to the recombination of excitons bound to impurities in direct-gap semiconductors [33].

The structure of the photoluminescence spectrum typical of high excitation rates (curve 4 in Fig. 8) was reported also in [27], where a comparison was made between the recombination radiation spectra of doped n- and p-type GaAs samples and the narrow luminescence lines were attributed to donor and acceptor impurities present in the semiconductor.

A high-sensitivity recording system enabled us, firstly, to determine the singularities in the photoluminescence spectra (beginning from a very low optical excitation rate of 6×10^{15} cm^{-2}·sec^{-1} right up to 10^{22} cm^{-2}·sec^{-1}) with a high spectral resolution and, secondly, to establish the relationship between the predominant recombination process and free excitons present in the semiconductor. This relationship could only be observed at relatively low optical excitation rates.

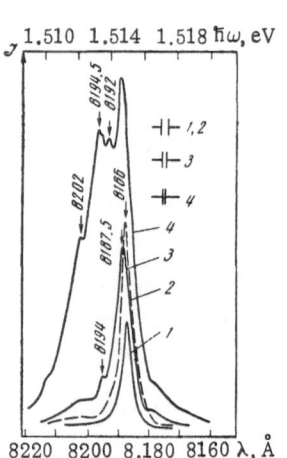

Fig. 8. Luminescence spectra of GaAs obtained at 2°K for different optical excitation rates (cm^{-2}·sec^{-1}): 1) 10^{16}; 2) 5×10^{17}; 3) 10^{18}; 4) 10^{21}.

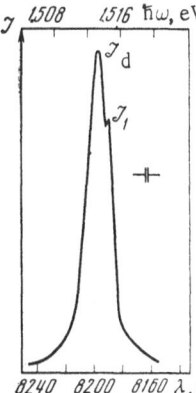

Fig. 9. Luminescence spectrum of a GaAs sample No. 4 obtained for $I \approx 5 \times 10^{21}$ cm$^{-2} \cdot$ sec^{-1}.

At helium temperatures the intensity of the free-exciton luminescence line at 1.5160 eV (8179 Å) was considerably weaker than the main luminescence lines. This made it extremely difficult to study directly the interaction between free excitons when their concentration was increased. For this reason some of the experimental results relating to this question were obtained during a study of the characteristics of the predominant luminescence.

At the maximum optical excitation rates which could be obtained by focusing the Ne—He laser radiation the structure of the spectrum began to disappear and instead a wide luminescence band $\mathcal{J}_d$ with an energy of 1.5135 eV was observed (Fig. 9).

§3. Photoluminescence of GaAs at Temperatures 2-100°K.

Investigation of Temperature Dependence of Recombination

Radiation Intensity

Since the energy of the thermal dissociation of localized excitons is considerably less than the binding energy of free excitons, a redistribution between the dissociation and binding processes may be expected when the temperature is varied.

Figure 10 shows the luminescence spectra of GaAs recorded at an optical excitation rate $I \approx 5 \times 10^{17}$ cm$^{-2} \cdot$ sec^{-1} at various temperatures. We shall show below that at this rate of optical excitation the dominant influence at 1.5145 eV is the luminescence $\mathcal{J}_1$ which is due to excitons bound to neutral donors. Figure 11a shows the temperature dependences of the intensity $\mathcal{J}_1$ of the exciton line, its half-width Δ, and wavelength. It is clear from this figure that at temperatures above that of liquid helium the intensity $\mathcal{J}_1$ rose rapidly and the line broadened. As expected, the emission wavelength varied with temperature in the same way as the wavelength λ corresponding to the forbidden-band width [20].

Fig. 10. Photoluminescence spectra of GaAs obtained for $I = 5 \times 10^{17}$ cm$^{-2} \cdot$ sec^{-1} at different temperatures (°K): 1) 10; 2) 26; 3) 32. The inset on the right shows the temperature dependence of the ratio of the luminescence intensity due to localized excitons $\mathcal{J}_1$ to the corresponding intensity of the luminescence of free excitons $\mathcal{J}_{ex}$.

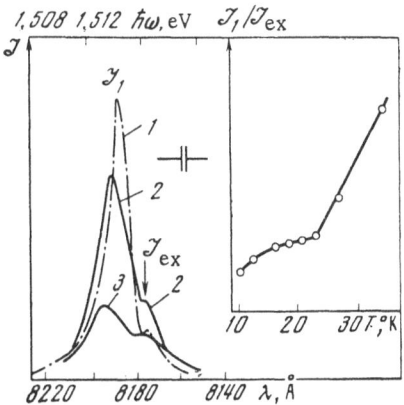

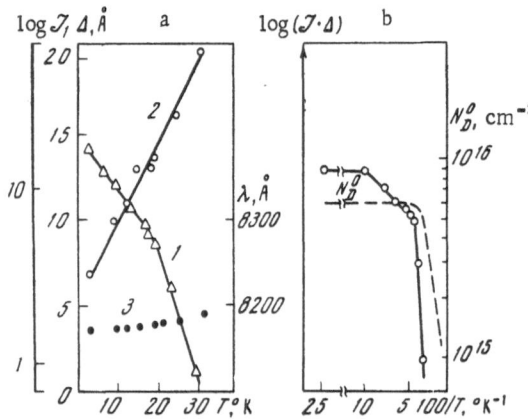

Fig. 11. Temperature dependences of the intensities of the luminescence lines due to excitons localized at neutral donors: a) intensity of the bound-exciton luminescence (1), its spectral half-width (2), and wavelength (3); b) integrated luminescence intensity $\mathcal{J}_1\Delta$. The dashed curve shows the temperature dependence of the neutral donor concentration N_D^0.

The integrated intensity of the bound-exciton luminescence decreased continuously with rising temperature (Fig. 11b). In the temperature range 10-20°K the fall was exponential and the thermal activation energy was $\zeta_1 = 0.86$ meV, whereas at T > 20°K this energy was $\zeta_2 = 9.5$ meV (Fig. 11b). This dependence of $\mathcal{J}_1$ on $1/T$ was in qualitative agreement with the definition of the fundamental luminescence as the radiation due to the recombination of excitons localized at neutral donors.

A quantitative calculation was difficult for the following reasons. The intensity of the luminescence due to excitons bound to neutral donors is proportional to the number of these donors N_D^0, number of free excitons n_{ex}, and a factor representing the thermal dissociation of the complexes $[D^0X]$:

$$\mathcal{J}_1 \propto N_D^0 n_a (1 - Ae^{-\zeta_1/kT}), \tag{12}$$

where A is a constant independent of the ionization (activation) energy ζ_1. This relationship reflects the law of mass action applied to the reaction of binding of an exciton to a neutral donor. At low temperatures, when the thermal ionization of neutral donors and excitons can be ignored, we have the relationship

$$\mathcal{J} \propto \mathcal{J}_0 (1 - Ae^{-\zeta_1/kT}).$$

For this reason the dependence observed experimentally in the range T < 20°K for all the investigated gallium arsenide samples was attributed to the dissociation of the $[D^0X]$ complexes. The energy of thermal dissociation ζ_1 was close to the theoretical value. At higher temperatures one should allow for the thermal ionization of donors and free excitons. It was quite easy to obtain a formula relating the concentration of neutral donors to temperature. In the hypothetic case of the existence of a single level in a semiconductor, the temperature dependence of the concentration of neutral donors is described by a dependence of the type [34]:

$$N_D^0 = N_D \left(1 - \frac{2}{1 + \sqrt{1 + 4\left(\frac{N_D}{\beta N_c}\right)\exp\frac{\mathcal{E}_D}{kT}}} \right), \tag{13}$$

where $N_c = 2[2\pi m_e(kT/h^2)]^{3/2}$; $\beta = 1/2$. The application of this formula is justified only in the range of relatively low temperatures and only when the number of electrons transferred to compensating impurity levels is small compared with the number of free carriers, i.e., $N_A \ll n_0 \ll N_D$. Figure 11b shows such a dependence for $N_D = 6 \times 10^{15}$ cm^{-3} and $\mathcal{E}_D = 6.4$ meV.

In the case of strong compensation of samples and when the condition $n_0 \ll N_A < N_D$ is satisfied, the value of $\ln N_D^0$ in the region of the exponential fall decreases twice as fast as the value calculated from Eq.(13) [34]. Unfortunately, the degree of compensation of our samples was not known.

The third factor, which influenced the dependence $\mathscr{J}_1$ (T), was the number of free excitons. Quantitative estimates of the exciton concentrations at various temperatures could not be obtained for a number of reasons, the most important of which was the absence of any reliable information on the lifetime of pairs and excitons generated in GaAs.

Thus, according to the above discussion we could only say that the first fall of $\mathscr{J}_1$ occurred due to the dissociation of the $[D^0X]$ complexes and the second, characterized by a thermal ionization energy of 9.5 meV, was largely due to the ionization of neutral donors. The two processes had the same result: the concentration of localized excitons decreased and the concentration of free excitons increased correspondingly. For this reason the fall of the intensity of the $\mathscr{J}_1$ line was accompanied by a relative rise of the free-exciton luminescence, whose intensity also fell rapidly with temperature (Fig. 10). Above 35°K neither the bound-exciton nor the free-exciton line was observed.

The restriction of the temperature range in which the short-wavelength luminescence was observed at low excitation rates was characteristic of all the investigated samples. A comparison of the experimental results obtained for different GaAs crystals indicated that the nature of the dominant luminescence was, to a considerable degree, governed by excitons localized at impurities.

At high optical excitation rates $(I > 10^{21}$ cm^{-2} · sec$^{-1})$ a wide band $\mathscr{J}_d$ appeared at 1.5135 eV on a background of narrow lines. The main features of the temperature dependences of the photoluminescence spectra observed at high optical generation rates were typical of all the investigated samples of GaAs, in spite of the differences in the spectral composition of the luminescence observed at low excitation rates and low temperatures. When the temperature was raised, the new wide band acquired a typical asymmetric profile with a more gently sloping short-wavelength (Fig. 12). The temperature dependence of the intensity of this luminescence did not exhibit a threshold in the investigated temperature range (Fig. 13). At T > 20°K we observed a variation of the emission wavelength similar to the variation of the wavelength λ corresponding to the forbidden-band width. The influence of rising temperature on the profile of the spectral curve, its half-width, and wavelength was typical of a system of nonequilibrium electrons and holes which was degenerate at low temperatures.

The maximum steady-state concentration of the generated pairs of nonequilibrium pairs N could be estimated roughly from the condition $N = IL_d^{-1}\tau$, where I is the optical excitation intensity, L_d is the diffusion length of the minority carriers, and τ is the nonequilibrium carrier lifetime. Measurements of the diffusion length of the minority carriers in GaAs at 77°K

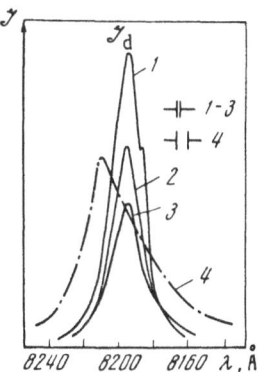

Fig. 12. Photoluminescence spectra of a GaAs sample No. 4 obtained for $I = 5 \times 10^{21}$ cm^{-2} · sec^{-1} at different temperatures (°K): 1) 4.2; 2) 20; 3) 35; 4) 60.

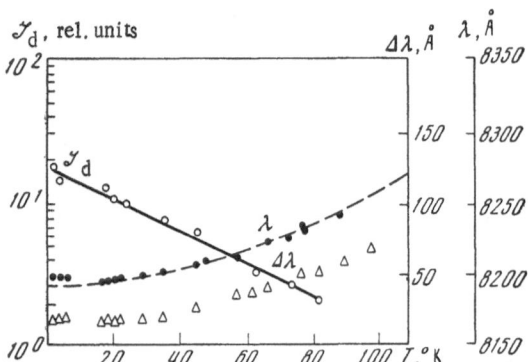

Fig. 13. Temperature dependences of the intensity of the photoluminescence due to electron-hole drops $\mathscr{J}_d$, its half-width $\Delta\lambda$, and wavelength at the maximum λ, plotted for a GaAs sample No. 4 excited at a rate $I \approx 5 \times 10^{21}$ cm$^{-2} \cdot$sec^{-1}. The dashed curve represents the temperature dependence of the wavelength corresponding to the forbidden-band width $\lambda_g(T)$ taken from [20].

[35] indicated that L_d was short: it amounted to 10^{-4}-3×10^{-4} cm. The nonequilibrium carrier lifetime τ at liquid-helium temperature was estimated experimentally from a luminescence pulse generated when gallium arsenide was excited with a Q-switched ruby laser. It was found that $\tau \leq 2 \times 10^{-8}$ sec for $I \approx 10^{21}$-10^{22} cm$^{-2} \cdot$sec^{-1}.

An estimate of the maximum density of nonequilibrium carriers generated at the rate of optical excitation corresponding to the appearance of the wide luminescence band gave $N \approx 10^{18}$ cm^{-3}, which was known to exceed the threshold corresponding to the onset of the degeneracy of both electrons and holes in GaAs.

The energy of the $\mathscr{J}_d$ luminescence was less than the forbidden-band width E_g. Thus, at $T = 2°K$, we had $\hbar\omega = 1.5135$ eV and $E_g = 1.5204$ eV [28]. For this reason and on the basis of all the other experimental observations we concluded that the appearance of this luminescence was due to the recombination of electrons and holes whose interaction altered the energy spectrum of the semiconductor.

§4. Photoluminescence Spectra of GaAs at T = 77°K

An investigation of the photoluminescence spectra of GaAs at liquid-nitrogen temperature gave in fact the information on the nature of the luminescence in the range 30-90°K. It was in this range of temperature T that the spectral half-width of the luminescence increased in proportion to T and the change in the energy of the luminescence maximum followed the corresponding change in the forbidden-band width of GaAs (Fig. 13). At 77°K when the optical excitation rate was $I \approx 10^{20}$-10^{21} cm$^{-2} \cdot$sec^{-1} only one luminescence line was observed (Fig. 14). Table 3 gives the principal parameters of the luminescence spectra of GaAs samples belonging to batches 1-4.

It is clear from Table 3 that the scatter of the energy positions of the luminescence maxima did not exceed $kT/2$ and the spectral half-width Δ was within the range $kT \leq \Delta \leq 2kT$.

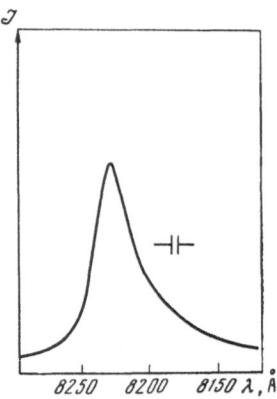

Fig. 14. Photoluminescence spectrum of GaAs at T = 77°K for $I = 10^{20}$-10^{21} cm$^{-2} \cdot$sec^{-1}.

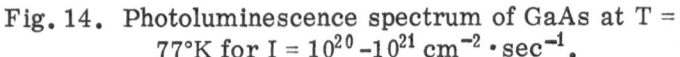

TABLE 3

GaAs batch No.	$\hbar\omega$, eV	λ, Å	$\Delta\hbar\omega$, meV	$\Delta\lambda$, Å
1	1.5075	8228	6.6	36
2	1.5089	8218	7.2	39
3	1.5067	8230	6.6	36
4	1.5058	8235	7.8	43

The luminescence spectra were investigated also in a wider range of photoexcitation rates when a ruby laser was used as the light source. When the optical excitation rate was increased, the observed line broadened continuously and, for example, for $I \approx 10^{23}$ cm$^{-2} \cdot$ sec^{-1} its half-width was $\Delta = 3\,kT$ (Fig. 15). In this case the photoluminescence spectrum did not have the characteristic form observed at low values of I. It should be noted that, in spite of the considerable spectral broadening of the luminescence, the maximum did not shift (within the limits of kT/2) along the energy scale.

When the rate of generation of nonequilibrium carriers by the Ne–He laser was within the range $I = 10^{20}$-10^{21} cm$^{-2} \cdot$ sec^{-1}, it was found that $\mathscr{J} \propto I^2$ (Fig. 16). Under these conditions the luminescence energy $\hbar\omega$ was practically constant. In the range $I > 10^{22}$ cm$^{-2} \cdot$ sec^{-1} the dependence of $\mathscr{J}$ on I was nearly linear.

§5. Discussion of Results

When the concentration nonequilibrium carriers in a semiconductor increases so that the separation between them becomes of the order of the Bohr exciton radius, the bound states of electrons and holes can no longer exist. The same conclusion (that individual excitons cannot exist) follows also if we allow for the Debye screening of the Coulomb interaction. The dissociation of excitons in the temperature range where the Maxwellian statistics is obeyed occurs at concentrations such that the Debye screening length

$$L_{\text{Deb}} = \left(\frac{kT\varkappa}{4\pi Ne^2}\right)^{1/2} \tag{14}$$

becomes less than the exciton radius a in the absence of screening. Estimates show that excitons should not exist at T = 77°K for a carrier concentration N = 5×10^{16} cm^{-3}. At low temperatures the Debye length L_{Deb} should be replaced with the Thomas–Fermi screening length $L_{\text{T-F}}$ [36]:

$$L_{\text{T-F}} = \left(\frac{F_{e,h}\varkappa}{6\pi Ne^2}\right)^{1/2}, \tag{15}$$

where $F_{e,h}$ is the Fermi energy.

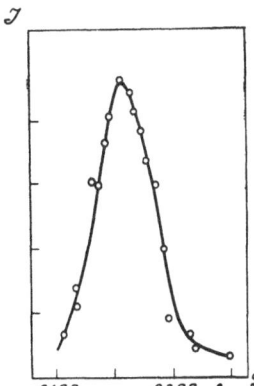

Fig. 15. Photoluminescence spectrum of GaAs at T = 77°K for $I \approx 10^{23}$ cm$^{-2} \cdot$ sec^{-1}.

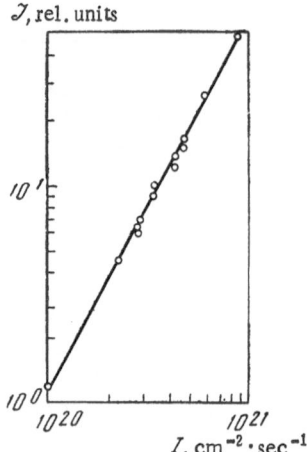

Fig. 16. Dependence of the luminescence intensity of GaAs on the optical excitation rate at $T = 77°k$.

The physical meaning of this substitution is that at low temperatures the degeneracy of free carriers occurs even at low concentrations. The concentration of particles in the allowed bands of semiconductors corresponding to the thermal degeneracy is given by

$$N^T_{e,\,h} = \frac{8\pi}{3}\left(\frac{2m_{e,\,h}}{h^2}\,kT\right)^{3/2}. \tag{16}$$

The Thomas—Fermi length $L_{T\text{-}F}$ is found to depend on the density of states at the Fermi surface. For this reason, in the most general case when $m_e \neq m_h$, the screening length in degenerate systems may be different for electrons and holes.

The relevant parameters for GaAs are collected in Table 4.

Thus, at $T \leq 4.2°K$ when the nonequilibrium carrier concentration is $N > 2.5 \times 10^{16}$ cm^{-3}, the electron—hole plasma is strongly degenerate.

When a degenerate Fermi gas is formed, the spin interaction between electrons and holes which are in allowed bands is found to be considerable compared with the kinetic energy of these particles. In the range $N > 10^{14}$ cm^{-3} the value of $r_s/a_{e,\,h}$, where r_s is the distance between neighboring particles and $a_{e,\,h}$ are the Bohr radii of electrons and holes, lies within a range in which we can use the results of calculations of the energy of collectively interacting electrons (or holes) in metals [36]. This interaction reduces the energy of electrons (holes) E by an amount governed by the sum of the exchange E_{exch} and correlation E_{corr} corrections so that E becomes

$$E = E_{kin} + E_{corr} + E_{exch}. \tag{17}$$

In the theory of metals the interaction energy of electrons (holes) is calculated on the assumption of the existence of a uniformly distributed compensating positive (negative) charge

TABLE 4

T, °K	N^e_{T}, cm^{-3}	N^h_{T}, cm^{-3}	N_{Deb}, cm^{-3}	$N^e_{T\text{-}F}$, cm^{-3}	$N^h_{T\text{-}F}$, cm^{-3}	Remarks
77	$6 \cdot 10^{16}$	10^{17}	$5 \cdot 10^{16}$	—	—	N_{Deb} calculated from $L_{Deb} = a = 110$ Å
4,2	$7,5 \cdot 10^{14}$	$1,3 \cdot 10^{15}$	—	$2,5 \cdot 10^{16}$	$4 \cdot 10^{15}$	$N_{T\text{-}F}$ calculated from $L_{T\text{-}F} = a = 110$ Å

or "background." Then, the exchange correction can be calculated in the Hartree—Fock approximation for electrons and holes in GaAs [36]

$$E_{exch}^{e,h} = -\frac{0.916}{r_s/a_0}\frac{R}{\varkappa}\ [eV],\tag{18}$$

where R is the Rydberg constant equal to 13.5 eV and a_0 is the Bohr radius of an electron in the hydrogen atom, which is 0.529×10^{-8} cm. In order to calculate the correlation correction to the total energy of electrons and holes, we can use the Wigner interpolation formula [36]

$$E_{corr}^{e,h} = -\frac{0.88}{r_s/a_{e,h}+7.8}\frac{m_{e,h}^*}{m\varkappa^2}R\ [eV],\tag{19}$$

where $m_{e,h}^*$ is the density-of-states effective mass of an electron and a hole.

The formula for the kinetic energy of electrons and holes can be expressed in the form

$$E_{kin}^{e,h} = \frac{2.21}{(r_s/a_0)^2}\frac{m}{m_{e,h}^*}R\ [eV].\tag{20}$$

Figures 17a-17c give the results of calculation of the interaction energies of electrons and holes and of the average energy per pair of particles. Since there is no rigorous analytic expression for the energy due to the interaction of carriers of opposite signs, these curves should not be regarded as absolutely accurate.

We can see from Fig. 17c that a system of electrons and holes has a stable energy minimum at $N = 3 \times 10^{16}$ cm^{-3}. At this concentration of free particles the excitons are not formed at low temperatures in GaAs and the system is completely degenerate. The minimum of the energy of the Fermi gas of electrons and holes in GaAs represents the state in which both holes and electrons are relatively stable. According to the calculations, the various components of the energy at this point are E_{exch} = -5.3 meV, E_{corr} = -2.3 meV, and E_{kin} = (3/5)(F_e + F_h) = 3.2 meV.

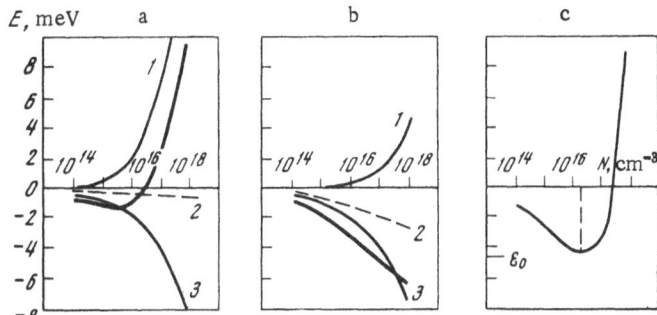

Fig. 17. Energy relationships calculated for electrons and holes interacting in GaAs: a) dependence of the energy of interacting electrons on their concentration; b) dependence of the energy of interacting holes on their concentration; c) dependence of the energy per electron-hole pair on the pair concentration; 1) E_{kin}; 2) E_{corr}; 3) E_{exch}. The continuous curves are the dependences of the energy E = E_{kin} + E_{exch} + E_{corr} on the carrier concentration.

At very low nonequilibrium carrier densities the Coulomb interaction also reduces the energy of electrons and holes and produces a discrete level $\mathcal{E}_0$ in the forbidden band of a semiconductor and this level represents the binding energy of excitons (Fig. 17c). Then, a comparison of the average energy per pair of particles at the energy minimum of a Fermi gas with the binding energy of excitons shows that the degeneracy of nonequilibrium carriers in GaAs may result from the dissociation of all excitons and formation of a system whose energy spectrum is affected by the collective interaction between particles. This model is in qualitative agreement with the experimental results obtained for different gallium arsenide crystals.

Calculations of the energy of the Fermi system in GaAs are valid at very low temperatures. Heating of this semiconductor increases the total energy of the system and causes a transition to a nondegenerate state. It is known that the distribution of the kinetic energies E_{kin} of carriers in the allowed bands of a nondegenerate semiconductor is described by a Maxwellian distribution function of the type $dN(E_{kin})/N(E_{kin}) \propto E_{kin}^{1/2} \exp(-E_{kin}/kT)\,dE_{kin}$. However, for $I = 10^{20}\text{-}10^{21}$ cm^{-1} at $T = 77°K$ the luminescence spectrum has a half-width Δ which is less than that predicted on the basis of the Maxwellian distribution (Table 4). In this range of optical excitation rates the radiative recombination of electrons and holes is of bimolecular nature (Fig. 16). The tendency for the narrowing of the luminescence spectrum of a nondegenerate electron—hole plasma in GaAs at $T = 77°K$ can be explained bearing in mind the details of the energy-band structure, i.e., that GaAs is a direct-gap semiconductor with valence and conduction bands of very different curvatures. In view of the absence of an inversion symmetry, the degeneracy of the heavy-hole bands is lifted in some crystals. It is known [37] that in the case of narrow-gap semiconductors the efficiency of radiative transitions involving light holes is higher than the efficiency of the corresponding transitions involving heavy holes. In this case the laws of conservation of energy and momentum may limit the interband radiative recombination. However, the available estimates of these effects for GaAs [37] fail to explain the experimental observations. Therefore, it is most likely that the observed narrowing of the photoluminescence spectra (Fig. 14) is due to the self-absorption of the luminescence emerging from a crystal.

The change in the luminescence line profile at high optical excitation rates may, to some extent, be explained by a change in the nature of the long-wavelength luminescence edge due to the narrowing of the forbidden band of the semiconductor with increasing nonequilibrium carrier density and by an increase in the temperature of the semiconductor and the appearance of a thermal heterojunction near the surface. Such a heterojunction may also result in reabsorption of the short-wavelength luminescence.

We shall conclude this section by pointing out that the experimental results and the calculated energy sheme do not exclude the possibility of the appearance of an intermediate state in which electron—hole drops are formed. In the situation discussed above the prime task is to explain the dependence of the binding energy of excitons $\mathcal{E}_0$ on their concentration n_{ex}. It is shown in [4] that an increase in n_{ex} should result in a corresponding reduction of $\mathcal{E}_0$. It was difficult to obtain experimental information on this point because the concentration of accidental impurities even in the best samples used in our study ($N_D = 6 \times 10^{15}$ cm^{-3}) was comparable with the density of nonequilibrium electrons and holes corresponding to the onset of their degeneracy at liquid-helium temperatures. For this reason the appearance of electron—hole drops in GaAs is not yet proven. In contrast to Ge and Si, electron—hole drops may be detected experimentally in a relatively narrow range of high exciton concentrations at very low temperatures.

§6. Supplement. Possibility of Existence of Condensate in Pure Epitaxial GaAs Films

In the last two years the interest in collective effects in semiconductors has risen considerably. Several new theoretical and experimental papers have appeared [38-44]. For

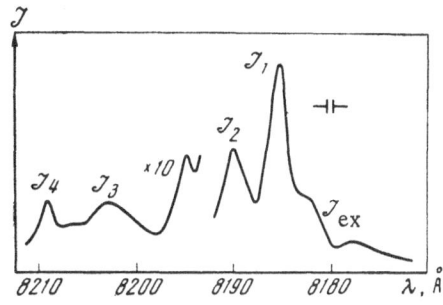

Fig. 18. Photoluminescence spectrum of GaAs
at T = 1.5°K for I ≈ 10^{17} cm^{-2}·sec^{-1}.

example, calculations of the condensate energy made allowing for the anisotropy and multivalley nature of semiconductor crystals are reported in [38]. It is shown that allowance for these factors favors the appearance of a condensed state. Calculations of the binding energy of exciton molecules [39, 40] have confirmed theoretical estimates [4]. Measurements of the scattering of light in matter [41, 42] and observation of luminescence oscillations in strong magnetic fields [43] have provided more reliable confirmations of the existence of electron-hole drops in Ge. As shown earlier, gallium arsenide is not a semiconductor in which it would be easy to observe experimentally the condensation of free excitons. The difficulty is that, in contrast to Ge and Si, it is usually a direct-gap material with allowed optical transitions and it is characterized by a fast ($\tau \leq 10^{-9}$ sec) relaxation of the concentrations of nonequilibrium carriers and excitons [45]. The lifetimes of free excitons in gallium arsenide may be shorter than the characteristic condensation times, which are known to be ≳10^{-8} sec in the case of Ge and Si [46, 47]. Consequently, the current ideas on the condensation mechanism may be inapplicable to gallium arsenide.

We shall describe here the experimental results which, nevertheless, demonstrate the existence of a condensed constant-density phase in gallium arsenide, which appears at low temperatures and high optical excitation rates.

The measurements were carried out on purer GaAs samples grown by the gas-transport epitaxy method. The maximum optical excitation rate of 2×10^{22} cm^{-2}·sec^{-1} was achieved by a sharp focusing of an argon laser with an output power of 160 mW. The samples were immersed in superfluid liquid helium at 1.5°K.

Figure 18 shows the photoluminescence of GaAs with n = 5×10^{13} cm^{-3} and μ = 135,000 cm^2·V^{-1}, sec^{-1}, measured at 77°K. It is clear from Fig. 18 that the spectra exhibit a series of well-resolved luminescence lines. The identification of some of these lines is given in several recent papers [45, 47, 49]. Nevertheless, a rigorous experimental proof* is still available only for one luminescence line ($\mathcal{J}_1$), which is due to the recombination of excitons localized at neutral donors [47]. All the other luminescence lines may be determined approximately from a calculated scheme of energy levels which appear in an excited semiconductor with shallow impurity centers [27]. According to the generally accepted ideas [27, 45, 47, 49], the luminescence band (line) $\mathcal{J}_{ex}$ is due to the recombination of free excitons. It has been shown theoretically [50] that the binding energy of free excitons calculated allowing for the actual band structure of GaAs does not differ greatly from that determined earlier [27] and it amounts to 4.2 meV. At low optical excitation rates (I < 10^{16} cm^{-2}·sec^{-1}) the probability of binding of carriers of opposite sign into excitons becomes less.

The photoluminescence spectra obtained under these conditions are dominated by the luminescence line $\mathcal{J}_2$ which is due to the recombination of free holes with shallow donors.

*In the reported photoluminescence spectra of pure epitaxial films [45-49] the luminescence energy $\mathcal{J}_1$ differs somewhat from one investigation to another and this is due to the deformation of the investigated samples.

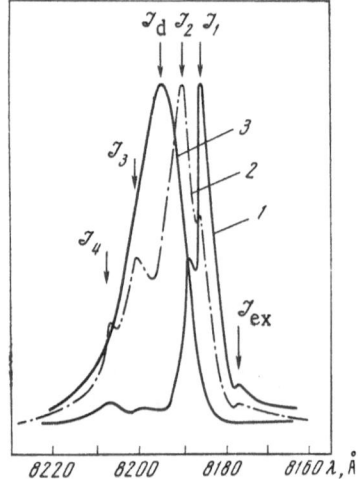

Fig. 19. Photoluminescence spectra of GaAs at T = 1.5°K for different optical excitation rates (cm^{-2} · sec^{-1}): 1) 5×10^{19}; 2) 2×10^{21}; 3) 5×10^{21}.

The ionization energy of these donors, deduced from the photoluminescence spectrum shown in Fig. 18, is 6.2 meV. The intensity of the long-wavelength components is considerably less than that of the main luminescence lines. The long-wavelength lines $\mathscr{J}_3$ and $\mathscr{J}_4$ are attributed in [47] to a complex recombination process involving excitons and shallow impurity centers.

Figure 19 shows the luminescence spectra obtained at various optical excitation rates for an epitaxial film with n = 5×10^{14} cm^{-3} and μ = 67×10^3 cm^2 · V^{-1} · sec^{-1}, measured at T = 77°K. It is found that in the range I > 10^{17} cm^{-2} · sec^{-1} the intensities of the lines $\mathscr{J}_{ex}$, $\mathscr{J}_1$, $\mathscr{J}_3$, and $\mathscr{J}_4$, which, according to [47], are of exciton origin and the intensity of the line $\mathscr{J}_2$ are all proportional to the optical excitation rate. At high excitation rates the short-wavelength components show a tendency to saturation. It is in this range that we observe a wide band $\mathscr{J}_d$, whose maximum coincides with the band described earlier in § 2 in the present chapter. The integrated intensity of $\mathscr{J}_d$ varies superlinearly with the excitation rate (as I^3) so that this band dominates the spectrum completely at excitation rates exceeding 3×10^{21} cm^{-2} · sec^{-1}. The main luminescence lines of epitaxial GaAs films were identified by recording the photoluminescence spectra in different electric fields applied to ohmic contacts. It was known that the degree of occupancy of the impurity centers and the cross section for the binding of carriers into excitons could be altered greatly by the application of an electric field capable of heating the majority carriers in a semiconductor. For example, it was reported in [15] that in such a semiconductor as Ge the breakdown of impurity centers, characterized by an avalanche ionization, was observed in relatively weak fields (w = 10 V/cm). A nonlinearity of the current—voltage characteristics was also noted in a study of the electrical properties of GaAs single crystals [51].

Figure 20 shows the dependences of the dark current J and of the photocurrent on the electric field w. The field was modulated at an audio frequency. The photocurrent signal picked up from a low-resistance load was amplified with a narrow-band amplifier and measured with a voltmeter. The modulation photoluminescence spectra were recorded by tuning the system to twice the repetition frequency. Curves 2, 3, and 4 were obtained at different optical excitation rates given in the caption of Fig. 20. In this particular sample the strong rise of the dark current, due to the breakdown of shallow donors, occurred at 3 V/cm. The electric field corresponding to the avalanche rise of the photocurrent depended on the illumination and at high illumination levels (I ≈ 5×10^{20} cm^{-2} · sec^{-1}) it was close to w = 1 V/cm.

Figure 21 shows two photoluminescence spectra recorded at I ≈ 5×10^{20} cm^{-2} · sec^{-1}. Spectrum 1 was obtained by modulation of nonequilibrium electrons and holes in a field of

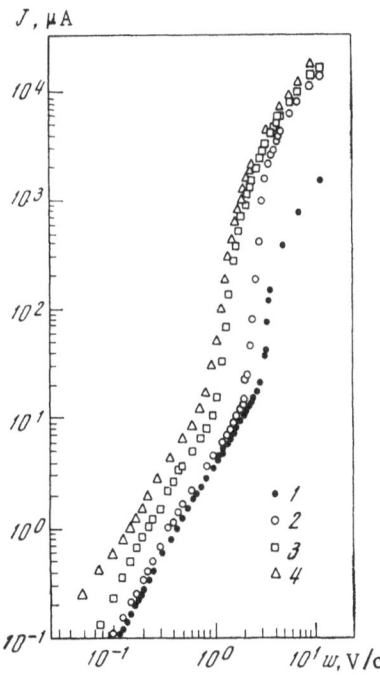

Fig. 20. Dependences of the dark current and photocurrent J on the electric field w: 1) in the absence of exciting radiation; 2) excited by scattered light; 3) illumination with an Ne–He laser (I ≈ 10^{18} cm^{-2} · sec^{-1}); 4) excited by illumination with Ar laser (I ≈ 5×10^{20} cm^{-2} · sec^{-1}).

w = 1.3 V/cm intensity, corresponding to a strong rise of the photocurrent (compare with curve 4 in Fig. 20). It is clear from Fig. 21 that all the components of exciton origin appeared in this case. These mechanically modulated spectra were observed only at the highest value of the photocurrent. The experimental observations were in agreement with the earlier interpretation of the $\mathscr{J}_1$ and $\mathscr{J}_2$ lines.

Characteristic features of the impact ionization in GaAs are, firstly, the presence of a fairly extended range in which the prebreakdown dependence $J \propto w^2$ is observed, and secondly, a dependence of the breakdown fields of excitons and impurity centers on the initial free-carrier density (Fig. 20). The latter explains largely why the measurements of the photocurrent and photoluminescence spectra in different electric fields failed to reveal a sharp boundary separating impurity and exciton breakdown mechanisms. The impact ionization in GaAs has been studied to a much lesser extent than, for example, in the case of Ge. The features of this ionization process deserve a separate study.

An electric field which causes impact ionization of free excitons and shallow donors does not affect the principal characteristics of the luminescence represented by the $\mathscr{J}_d$ band. In the presence of a static electric field the $\mathscr{J}_d$ band is observed also at relatively low optical excitation rates whereas in the absence of a field the spectrum is dominated by the lines due to free and bound excitons. The position and profile of the $\mathscr{J}_d$ band remain unchanged

Fig. 21. Photoluminescence spectra of GaAs modulated by the application of an electric field w (V/cm): 1) 1.3; 2) > 2.

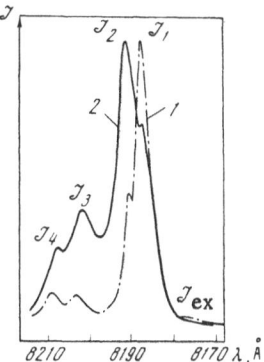

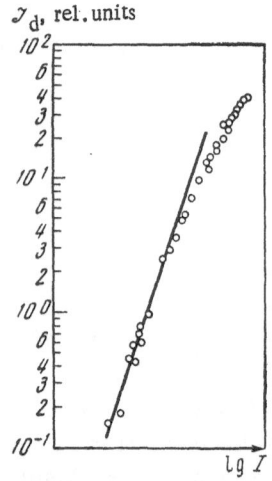

Fig. 22. Dependence of the intensity of the drop luminescence $\mathscr{J}_d$ on the optical excitation rate I in a field w = 4 V/cm.

compared with the corresponding characteristics at the highest excitation rates. Hence, it is possible to determine the dependence of the integrated luminescence intensity $\mathscr{J}_d$ on the excitation rate in a wider range of the latter (Fig. 22). The experimental points plotted in this figure were obtained by applying an electric field w = 4 V/cm, which suppressed completely the processes in which excitons and shallow impurity centers participated. The available experimental data suggested that the $\mathscr{J}_d$ luminescence band could be due to "metallic" electron–hole drops. However, in contrast to germanium and silicon, free electrons and holes and not excitons condense in gallium arsenide, as indicated by the above results obtained in electric fields.

We calculated the profile of the luminescence line of a degenerate electron–hole plasma in semiconductors with direct allowed transitions in order to find the density of the condensate N from the experimental data. The profile was calculated employing the formula

$$\mathscr{J}(\hbar\omega - E_d') \propto \frac{(\hbar\omega - E_{gd})^{1/2}}{\left[\exp\dfrac{(\hbar\omega - E_{gd})\frac{\mu}{m_e} - F_e}{kT} + 1\right]\left[\exp\dfrac{(\hbar\omega - E_{gd})\frac{\mu}{m_h} - F_h}{kT} + 1\right]},\tag{21}$$

where E_{gd} is the forbidden-band width for carriers recombining in drops.

The calculated and experimental curves were made to coincide at the position of the maximum. The best agreement between theory and experiment was obtained for an equilibrium concentration $N = 1.1 \times 10^{16}$ cm^{-3} at T = 3.8°K (Fig. 23).

Estimates of the principal characteristics of electron–hole drops in gallium arsenide, given in § 5 in the present chapter, yield $N = 3 \cdot 10^{16}$ cm^{-3}. The discrepancy between this value

Fig. 23. Photoluminescence spectra of GaAs: 1) experimental curve for $I = 10^{22}$ cm$^{-2} \cdot$ sec^{-1} at T = 1.5°K; 2) curve calculated using Eq. (21) for $N = 1.1 \times 10^{16}$ cm^{-3} at T = 3.8°K.

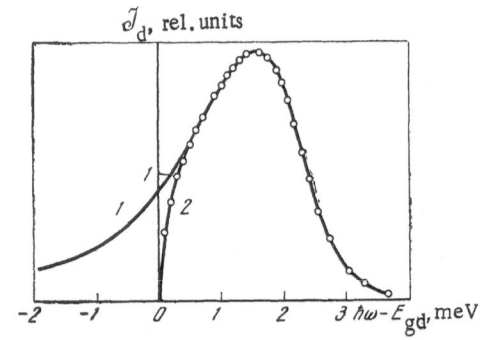

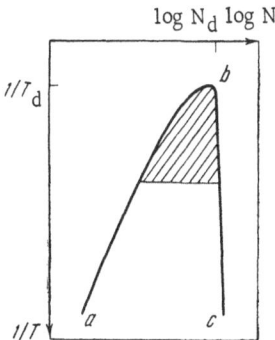

Fig. 24. Gas—liquid phase diagram (curve abc).

and the one obtained from the luminescence spectra is due to the existence of "cross" transitions associated with the Auger recombination of electrons and holes in drops. This Auger recombination undoubtedly takes place, as indicated by the long-wavelength tail of the experimental curve (Fig. 23).

The calculated local change in the forbidden-band width due to the correlation and exchange interaction of carriers is 7.6 meV. The depth of the potential well deduced from the long-wavelength luminescence edge is 8.7 meV, which is in satisfactory agreement with the calculated value. The slight discrepancy is not unexpected because numerical estimates of the condensate energy are obtained ignoring the correction due to the correlation interaction of carriers of opposite sign.

In the case of an electron—hole condensate the phase diagram can be represented, by analogy with a liquid—gas system, by a curve of the type designated abc in Fig. 24. As shown earlier, under normal conditions the photoluminescence spectra of GaAs show evidence of electron—hole drops only at high optical excitation rates. The spreading of the short-wavelength edge of the luminescence band of the condensate (Fig. 23) is evidence of a considerable overheating of the active region in the sample. Consequently, measurements of the photoluminescence of the condensate reveal only a small part of the phase diagram near the critical point (shown shaded in Fig. 24). This means that, under the experimental conditions, the densities of particles in both phases are comparable. This gives rise to considerable difficulties in the determination of the temperature dependence of the drop luminescence intensity. This may explain why the earlier measurements (Fig. 13) failed to reveal the range of existence of the condensed phase along the temperature scale. However, the anomalous behavior of the points representing the energy of the maximum of this luminescence at low temperatures may be regarded as evidence that electron—hole drops are observed even in these experiments at helium temperatures.

CHAPTER IV

CHANGE IN ABSORPTION COEFFICIENT OF UNDOPED GaAs DUE TO STRONG OPTICAL EXCITATION

Strong optical and electron excitation may give rise to laser action in undoped GaAs crystals [52-54]. The stimulated emission maximum at 77 and 300°K is less than the corresponding forbidden-band width. It is suggested in [53] that this may be due to the narrowing of the forbidden band as a result of screening of the crystal field by free carriers and because of interaction of carriers with one another. These two effects occur also in systems of electrons and holes close to degeneracy [55].

The energy position of the interband luminescence maximum may be determined, in principle, if we know the potential energy of a system, governed by the interaction between nonequilibrium carriers, and the kinetic energy. In degenerate systems the kinetic energy is governed by the positions of the quasi-Fermi levels of electrons and holes, whereas in nondegenerate systems it is governed by the thermal motion of free particles. It follows that the variation of the forbidden-band width of a nondegenerate semiconductor can be deduced from the nature of the variation of the long-wavelength edge of the luminescence due to nonequilibrium carriers. The contribution of low-energy photons may be particularly large at high nonequilibrium carrier concentrations when the probability of radiative transitions and energy losses rise at the same time. Since all the known calculations of systems of interacting carriers are approximate, it would be interesting to carry out an experimental investigation. In our study, effects of this kind were observed in the spectra of the photoluminescence of GaAs emitted from the surface of a sample opposite to the illuminated face. Experimental results also indicated violation of the classical Bouguer law, predicting an exponential fall of the luminescence in solids.

Undoped GaAs grown from the melt had the properties listed in Table 2. The investigation was carried out on plane-parallel plates polished on both sides but not given any resonator structure. Since the wavelength of the ruby laser radiation was 6943 Å, this radiation was absorbed in a thin surface layer ($d = 1\ \mu$). The recombination radiation generated at a depth equal to the diffusion length of nonequilibrium carriers was recorded when it emerged from the crystal on the opposite side.

Near the fundamental absorption edge the excited face of a sample emitted one luminescence band. When the incident photon flux reached 10^{22}–10^{23} cm^{-2} · sec^{-1}, the intensity maximum of this band at T = 77°K was located near λ = 8230 Å ($\hbar\omega$ = 1.5075 eV) (Fig. 15) and at T = 4.2°K it was close to λ = 8197 Å ($\hbar\omega$ = 1.5135 eV) (Fig. 7). A typical spectrum of the luminescence which traversed a sample 90 μ thick is shown in Fig. 25 (curve 1). This spectrum has a pronounced asymmetry with a steep short-wavelength edge and gentler slope on the long-wavelength side. For comparison, the same figure includes also the luminescence spectrum of GaAs observed on the excited face side (curve 2). In both cases the measurements were carried out at T = 77°K using a photon flux $\geq 5 \times 10^{21}$ cm^{-2} · sec^{-1}. It is clear from Fig. 25 that the separation between the maxima was about 100 Å (0.015 eV). The intensity of the luminescence which traversed the sample was a nonlinear function of the excitation rate. The results obtained for different rates in the ratio 2 : 3 : 8 are represented by curves 1-3 in Fig. 26. The amplitude of curve 3 in Fig. 26 was reduced by a factor of 5 to fit the figure. Thus, when the excitation rate was increased by a factor of 4 (between curves 1 and 3), the luminescence intensity rose by an order of magnitude.

The shift of the luminescence intensity maximum in the direction of longer wavelengths as a result of traversal of the luminescence across the sample was not unexpected and it was due to the absorption of the luminescence in the bulk of the crystal. However, the dependences of the amplitude and profile of the luminescence band on the thickness of the sample were un-

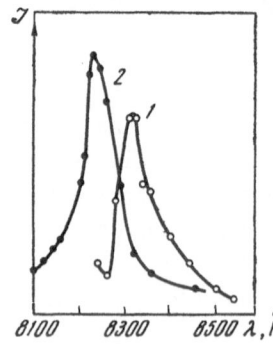

Fig. 25. Luminescence spectra of undoped GaAs at 77°K: 1) luminescence which crossed a sample 90 μ thick excited at a rate I $\approx 8 \times 10^{21}$ cm^{-2} · sec^{-1}; 2) luminescence observed on the illuminated side when a sample was excited at a rate I $\approx 5 \times 10^{21}$ cm^{-2} · sec^{-1}.

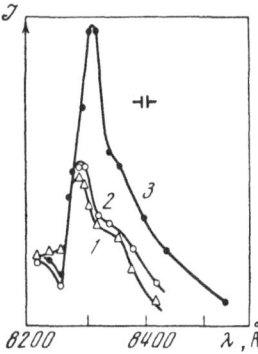

Fig. 26. Spectra of the luminescence which crossed a
sample of undoped GaAs 90 μ thick at 77°K. The explanations are given in text.

usual. Figure 27a shows the spectra of the luminescence which traversed samples 100 and
600 μ thick at liquid-nitrogen temperature. The measurements were carried out at two different photoexcitation rates. For a given excitation rate the luminescence which crossed the
600-μ sample shifted in the direction of longer wavelengths by less than 20 Å relative to the
luminescence which crossed the 100-μ sample. The maximum intensity of the luminescence
decreased only moderately (by a factor exceeding 2). It is clear from Fig. 27 that the photoluminescence curves 1 and 2 were considerably broadened compared with curves 3 and 4 because
of the slow fall in the long-wavelength range.

Similar measurements were carried out also at T = 8°K. Figure 27b shows the luminescence spectra obtained for the same samples and at the same photoexcitation rates as in the
preceding experiment. The maxima of the luminescence curves were now located at λ = 8280-
8290 Å. The change in the positions of the luminescence maxima as a result of cooling was
in agreement with the temperature-induced change in the forbidden-band width. These curves
differed from those plotted in Fig. 27a by a narrowing of the luminescence bands.

In the case of thick samples (d $\approx$ 0.1 cm) the luminescence signal which crossed a sample
was comparable with the background signal due to the scattered light.

One of the unexplained effects was the weak dependence of the position of the maximum
of the luminescence that crossed a sample on its thickness for a fixed excitation rate. This

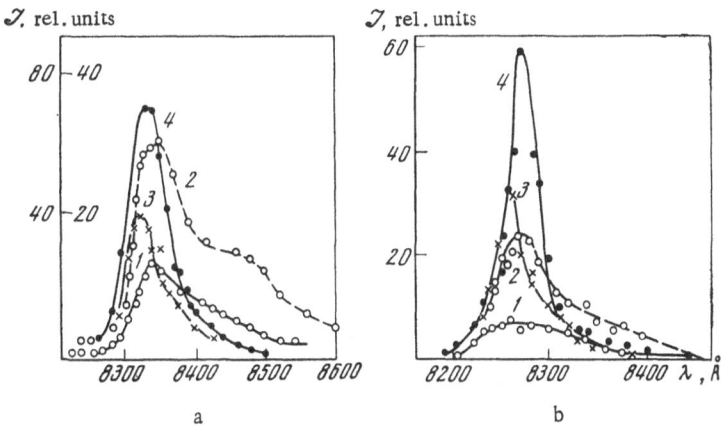

Fig. 27. Spectra of the luminescence which crossed samples 100 μ (3, 4) and 600 μ (1, 2) thick at 77°K (a) and at
8°K (b). Excitation rate (cm$^{-2} \cdot$ sec^{-1}): 1, 3) 10^{23}; 2, 4)
2 $\times$ 10^{23}. The left-hand scale applies to curves 3 and 4 and
the right-hand scale applies to curves 1 and 2.

behavior was a violation of the Bouguer law, which predicted an exponential attenuation of radiation in solids. Calculations indicated that when the Bouguer law was satisfied, the separation between the maxima in Fig. 27 should exceed 80-100 Å [56]. The absorption in the range $\lambda <$ 100 cm^{-1} was known to represent a sum of the fundamental and impurity absorption [20]. Since the contribution of the impurity absorption was limited by the number of impurity centers and the degree of their occupancy, a saturation of the impurity absorption should occur at some optical excitation rate and all the excess photons should cross a semiconductor attenuated to a much smaller degree, which would represent an effective bleaching of a semiconductor. At T = 77°K in the case of samples 80-100 μ thick the excitation rate at which such bleaching began was of the order of 5×10^{21} cm$^{-2} \cdot$sec^{-1} (Fig. 26).

A characteristic feature of the luminescence which crossed a crystal was the dependence of the energy of its maximum on the optical excitation rate I. For example, at T = 77°K the shift of the luminescence maximum in the direction of longer wavelengths was 11 meV for I = 5×10^{21}-10^{23} cm$^{-2} \cdot$sec^{-1}. This shift was far too large to be attributed simply to the heating of the semiconductor.

The observed change in the absorption coefficient of GaAs could also result in a shift of $\hbar\omega$ but in the opposite direction. For this reason the most likely explanation of the above experimental observation is the assumption of the effective narrowing of the forbidden band of the semiconductor at high nonequilibrium carrier densities. The reported results should be regarded as a qualitative confirmation that phenomena of this kind occur in semiconductors.

CHAPTER V

INVESTIGATION OF PHOTOLUMINESCENCE SPECTRA OF SILICON AT DIFFERENT OPTICAL EXCITATION RATES

§1. Review of Literature

Silicon is a covalent semiconductor with noncoincident extrema in the **k** space. It is known that the edge of the valence band of Si is located at the center of the Brillouin zone and the corresponding edges of the conduction band are in six equivalent crystallographic positions along the ⟨100⟩ axis. In the case of semiconductors with extrema located at different points in the **k** space we can have optical transitions which create electron–hole pairs with zero or nonzero quasimomentum, the latter being governed by the structure of the semiconductor. The laws of conservation of energy and momentum determine the relationships revealed by experimental studies of absorption and luminescence processes. For example, indirect excitons may appear in the luminescence and absorption processes only in a dissipative system, such as a crystal containing imperfections and impurity centers or subject to lattice vibrations. Therefore, the positions of the exciton lines in the luminescence or absorption spectra are not simply governed by the binding energy but are subject to a shift in one or the other direction due to the presence of a third particle.

In spite of the existence of extensive experimental information on the structure of the energy bands of Si, the forbidden-band width of silicon has not yet been determined sufficiently accurately. This indeterminacy in the values of E given by different authors is due to the corresponding indeterminacy of the binding energy of free excitons $\mathcal{E}_0$. Excitons were first discovered in silicon in a study of the absorption of light [57]. The binding energy of excitons was calculated using the effective-mass approximation in the case of simple valence and conduction bands [58]. Semiconductors with a degenerate valence band and with extrema located at different points in the quasimomentum space were considered in [59]. The binding energy of indirect free excitons was determined in [60] making allowance for the six equivalent minima in the conduction band and for their anisotropy. Allowance was also made for the contribution of the valence band split off from the degenerate bands by the spin–orbit interaction. The calculated binding energy of indirect excitons in silicon was $\mathscr{E}_0 = 12$ meV and the radius of the first unexcited orbit was found to be $a = 40$ Å.

In the indirect (phonon-assisted) transition range the exciton absorption lines were not discrete and this made it difficult to determine the binding energy of free excitons directly from the absorption spectra. The value of $\mathscr{E}_0$ estimated from indirect absorption transitions [57, 61] was 0.01 eV. The use of a more sensitive differential (modulation) method for the recording of the radiation transmitted by crystals made it possible to determine more accurately [62] the binding energy of indirect excitons in silicon $\mathscr{E}_0 = 14.7$ meV as well as the forbidden-band width $E_g = 1.17$ eV at $T = 1.8°$K.

Since the gap between the two conduction-band minima located at $\mathbf{k} = 0$ and at the point corresponding to the conduction band extremum was large ($\Delta E = 1.5$ eV), the occupancy of the six equivalent minima with nonequilibrium carriers was practically negligible. For this reason the recombination processes were governed by the absolute extrema of the energy band of silicon.

A considerable amount of work on the luminescence of silicon was carried out by Haynes and his colleagues [63-65]. The use of optical excitation enabled them to determine the nature of the luminescence lines of specially doped semiconductors [65] and to attribute them unambigously to impurities present in silicon. In the case of undoped samples Haynes et al. were the first to observe the free-exciton luminescence at 18 and 77°K [63]. The line profile was found to be of the Boltzmann nature. A slight deviation from the Boltzmann distribution, indicated by the luminescence line broadening, was attributed to the interaction between excitons and the crystal lattice. The absorption data [57] and the principle of detailed equilibrium [66] were used in a calculation of the luminescence spectra of indirect excitons in silicon and these were compared with the experimental results in order to determine the binding energy of free excitons, which was 0.008 eV.

At $T = 18°$K the main luminescence line due to the recombination of free excitons accompanied by the emission of transverse optical phonons had a maximum at 1.096 eV. Cooling revealed two closely spaced luminescence lines of comparable intensity at 1.0956 and 1.0974 eV [67]. It was also suggested there that the observed doublet was due to the splitting of the ground exciton state as a result of the valley–orbit interaction. A comparison of the low-temperature luminescence and absorption spectra yielded $\mathscr{E}_0 = 11.5$ meV and $E_g = 1.1645$ eV. The inconsistency of the valley–orbit splitting model was demonstrate in [62]. It was found that the luminescence lines reported in [67] were mirror images of the absorption lines due to indirect free excitons accompanied by the emission of transverse longitudinal optical phonons.

Thus, in spite of the very large experimental material available on this classical semiconductor, the binding energy of free excitons in silicon and the forbidden-band width of this material were not determined sufficiently accurately.

Further studies of the photoluminescence spectra were reported in [68, 69, 70-72]. This series of papers [68-72] was stimulated by the discovery of a low-temperature luminescence line at 1.083 eV, which appeared at high excitation rates [64]. The width of this luminescence line was considerably greater than kT and the luminescence intensity increased as the square of the intensity of the exciton line. A phenomenological physical description of this luminescence was given by Haynes in [64]. According to this description, exciton molecules formed at low temperatures and sufficiently high carrier densities and these molecules were destroyed by a form of the Auger process. One of a pair of bound excitons recombined producing a phonon and gave some of its energy to the remaining exciton. This exciton dissociated producing a free electron and a free hole, which had some kinetic energy. The principal conclusion of

Haynes's analysis was that the agreement between the experimental data and the recombination of exciton molecules was obtained only if it was assumed that the binding energy of free excitons in silicon was $\mathcal{E}_0 = 8$ meV.

An explanation of the experimental results based on the existence of exciton molecules [64] was used also in [68, 69]. The solution of the system of rate equations for particles participating in recombination was used in [68] to show that at low optical excitation rates I the concentrations of excitons n_{ex} and of exciton molecules (biexcitons) n_B should vary with I in accordance with the laws $n_{ex} \propto I$ and $n_B \propto I^2$, whereas at high optical excitation rates these concentrations should vary in accordance with the laws $n_{ex} \propto I$ and $n_B \propto I$. These dependences confirmed the reasoning given above. It should be noted that in the investigation reported in [68] the nonequilibrium carriers were generated either by a high-power gas laser or by electron bombardment. An electron accelerator, capable of producing short (86 nsec) pulses, was used in [69]. This made it possible to study the kinetics of recombination processes responsible for the luminescence lines at 1.098 and 1.083 eV observed at low temperatures. At T = 1.7°K the long-wavelength luminescence decayed in a time $\tau = 143$ nsec. When the temperature was increased, the decay time increased slightly and became $\tau = 155$ nsec at T = 4.2°K. At temperatures such that the luminescence with $\hbar\omega = 1.083$ eV was also observed the time dependence of the free-exciton luminescence had an initial region characterized by a fast decay of the luminescence intensity followed by a slower exponential decay. Measurements of the exciton lifetime at T = 20°K, when the long-wavelength luminescence was not observed, gave a value of 2.6 μsec.

The experimental investigations reported in [70-72] were stimulated by the prediction that an exciton gas in a semiconductor may condense into a liquid phase composed of electron–hole drops with properties resembling a metal rather than a dielectric liquid [3]. A qualitative consequence of this theoretical prediction is that phase transitions of this kind are highly likely in semiconductors with relatively large binding energies of long-lived free excitons. At first sight, it would seem that such a classical semiconductor as Si would be an ideal material because indirect excitons with long radiative lifetimes and a binding energy corresponding to an equivalent temperature much higher than that of liquid helium have been observed in this semiconductor.

It was suggested in [70-72] that the appearance of the long-wavelength luminescence at $\hbar\omega = 1.083$ eV was due to the formation, at suitable excitation rates and low temperatures, of electron–hole drops in silicon. In contrast to the results reported in [68, 69], it was found in [70] that at low excitation rates the intensity of the long-wavelength luminescence varied proportionally to I^3, whereas at higher photoexcitation rates this dependence was weaker. The rate equation was solved on the assumption that electron–hole drops were formed in the bulk and it was found that the number of particles in these drops increased proportionally to the cube of the optical excitation rate when the number of condensation centers was independent of I. The hypothesis of a degenerate electron–hole gas in the drops was used in the calculation of the expected luminescence line profile, which agreed well with the experimental observations obtained when the concentration of electron–hole pairs was $N = 2 \times 10^{18}$ cm^{-3}.

In the study reported below the presence of electron–hole drops in Si was deduced from the photoconductivity observed at different excitation rates and in different fields. Moreover, we investigated the luminescence spectra of Si beginning from the moment of appearance of a new line right up to the highest rates of optical excitation provided by a Q-switched ruby laser. We studied the photoluminescence spectra of crystals with different impurity concentrations. We also made quantitative measurements at various temperatures in order to determine the thermal threshold of the disappearance of the new luminescence and its relationship to the presence of free excitons in Si.

§ 2. Experimental Investigation of the Photoluminescence

of Si at Different Optical Excitation Rates

In this section we shall report the experimental results obtained in a study of the photoluminescence spectra of Si under continuous and pulsed optical excitation conditions. The measurements were carried out on various undoped p-type crystals whose resistivity at T = 300°K was ρ = 350-10^4 Ω·cm and in which the nonequilibrium carrier lifetime was τ = 100-1000 μsec.

The investigated samples were rectangular plates of various thicknesses. All the surfaces were subjected to a fine polishing and before each measurement a sample was etched for 1-2 min in a composite etchant (CP-4) and then washed in boiling distilled water for 20-30 min.

The photoluminescence spectrum of p-type Si obtained at an excitation rate I = 6 × 10^{18} cm^{-2}·sec^{-1} at T = 4.2°K is shown in Fig. 28. We observed two main luminescence peaks at 1.083 and 1.092 eV. At this optical excitation rate the luminescence at 1.083 eV was stepped because of the presence of unknown luminescence lines at 1.086 and 1.080 eV. The nature of the spectra was the same for different p-type crystals with room-temperature resistivities ρ = 100-450 Ω·cm. Only the ratios of the intensities of the main luminescence lines varied somewhat from sample to sample. In the case of Si crystals subjected to special purification treatment and characterized by a room-temperature resistivity ρ = $10^4 \Omega$·cm, we observed only the luminescence at 1.083 and 1.092 eV. All the other luminescence lines were clearly due to some impurities contained in the samples of Si with lower resistivities. A series of equidistant well-resolved luminescence lines was observed in [70]. These were attributed to complexes composed of several excitons captured by impurity centers. The luminescence at 1.092 eV was typical of all the p-type crystals and it was due to the recombination of an exciton bound to a neutral boron atom, accompanied by the emission of a transverse optical phonon[65].

The wide luminescence band at 1.083 eV was first observed by Haynes [64] and it was regarded by him as the luminescence due to the recombination of exciton molecules. This is not the view of the present author who is of the opinion that this band is due to the recombination of excitons condensed into a liquid phase composed of electron—hole drops.

When silicon was excited with high-power pulses corresponding to I = 10^{21} cm^{-2}·sec^{-1} the photoluminescence spectra of all the investigated crystals included not only the dominant band at 1.083 eV, but also an additional band at E = 1.098 eV, which was due to the recombination of free excitons (Fig. 29).

Since the concentration of bound excitons was limited by the number of impurities present in the semiconductor, the intensity of the 1.092 eV luminescence was negligible compared with

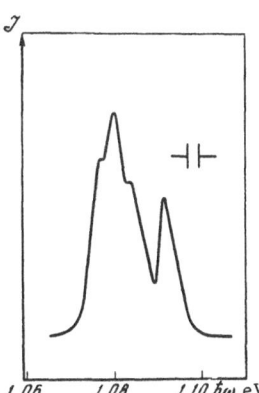

Fig. 28. Photoluminescence spectrum of Si (ρ = 350 Ω·cm at T = 300°K), recorded at T = 4.2°K for I = 6 × 10^{18} cm^{-2}· sec^{-1}.

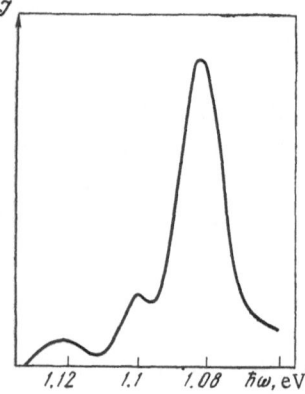

Fig. 29. Photoluminescence spectrum of Si ($\rho = 350$ $\Omega \cdot$ cm at T = 300°K), recorded at T = 4.2°K for I = 6×10^{22} cm$^{-2} \cdot$ sec^{-1}.

Fig. 30. Photoluminescence spectra of Si ($\rho = 450$ $\Omega \cdot$ cm at T = 300°K), obtained at different optical excitation rates (cm$^{-2} \cdot$ sec^{-1}): 1) 6×10^{18}; 2) 3×10^{18}; 3) 10^{18}.

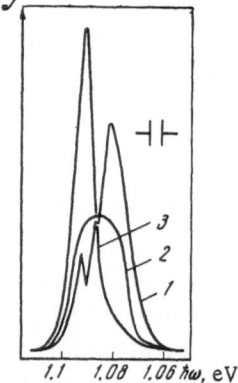

the considerably stronger luminescence at 1.083 eV. In the short-wavelength part of the spectrum we observed a wide luminescence band whose maximum was separated by 43 meV from the main band. The energy of this luminescence corresponded to a recombination process analogous to the dominant luminescence but with the particpation of acoustic phonos [73].

When the photoexcitation intensity was I < 6×10^{18} cm$^{-2} \cdot$ sec^{-1}, the relative intensity of the bound-exciton luminescence band at 1.092 eV was higher (Fig. 30). The long-wavelength luminescence was not observed even in the purest samples when the excitation rate was I $\approx 10^{18}$ cm$^{-2} \cdot$ sec^{-1}. Thus, all the major changes which occurred in the luminescence spectra of silicon containing impurities in concentrations not exceeding 10^{13} cm^{-3} were governed primarily by the optical excitation rate.

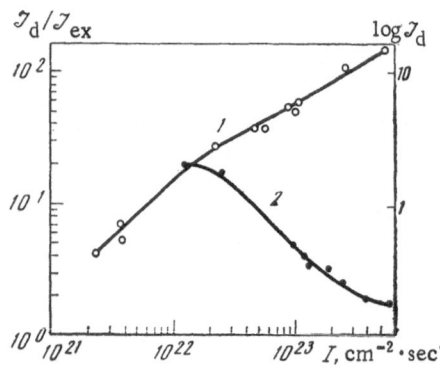

Fig. 31. Dependences of the intensity of the drop luminescence $\mathcal{J}_d$ at 1.083 eV (1) and of the ratio of this intensity to the intensity of the free-exciton luminescence $\mathcal{J}_{ex}$ (2) on the optical excitation rate.

A study of the drop (condensate) luminescence intensity $\mathcal{J}_d$ as a function of the optical excitation rate I [70] during continuous illumination with a laser radiation of up to 160 mW power revealed a region with a cubic dependence of $\mathcal{J}_d$ on I. In the present investigation it was found that when I was varied within the range 10^{21}–10^{22} cm$^{-2} \cdot$ sec^{-1} the intensity of the luminescence emitted by electron—hole drops rose linearly with the excitation rate. At high excitation rates this rise slowed down and a free-exciton luminescence line appeared in the spectrum (Fig. 31). The dependence of $\mathcal{J}_d/\mathcal{J}_{ex}$ on the excitation rate I is plotted in Fig. 31.

§3. Photoluminescence Spectra of Si at Different Temperatures. Investigation of the Temperature Dependence of the Luminescence Intensity

Both hypotheses put forward to explain the 1.083 eV luminescence postulated that this luminescence was due to the presence of a sufficiently large number of free excitons in the semiconductor. The experimental results indicated that at liquid-helium temperature and for an excitation rate I $\approx 6 \times 10^{18}$ cm$^{-2} \cdot$ sec^{-1} the photoluminescence spectrum of silicon was dominated by the 1.083 eV band and the luminescence due to the direct recombination of free excitons was not observed. The free-exciton line appeared only at considerably higher excitation rates. The cause of these changes in the spectra with the excitation rate was sought in the results of a study of the photoluminescence of Si at different temperatures.

The photoluminescence spectra changed with temperature. The intensities of the 1.083 eV and bound-exciton bands decreased continuously with rising temperature. At T > 10°K the photoluminescence spectrum of Si obtained for I = 6×10^{18} cm$^{-2} \cdot$ sec^{-1} did not include the 1.092 eV band (Fig. 32). At T > 4.2°K we observed the free-exciton line whose intensity rose strongly with temperature, reaching its maximum at T = 22°K. Only the free-exciton luminescence was observed above this temperature. Figure 33 shows the temperature dependences of the intensities of the free-exciton and drop (condensate) luminescence lines. At the beginning of a thermal cycle (2-14°K) the intensity of the E = 1.083 eV luminescence fell somewhat more slowly than at T > 12°K. The temperature at which this quenching occurred was equal to the temperature corresponding to the maximum intensity of the free-exciton luminescence (Fig. 33). The temperature dependence of the exciton luminescence intensity plotted on a semilogrgarithmic scale (Fig. 33) exhibited two linear regions with different slopes.

At low optical excitation rates the photoluminescence spectra were dominated by the recombination of the excitons captured by boron atoms. The appearance of free excitons was observed only when the temperature was raised (Fig. 34). It is clear from Fig. 34 that the intensity of the bound-exciton luminescence obtained when the excitation rate was I $\approx 5 \times 10^{17}$ cm$^{-2} \cdot$ sec^{-1} fell strongly near T = 10°K, whereas the relative intensity of the free-exciton luminescence increased.

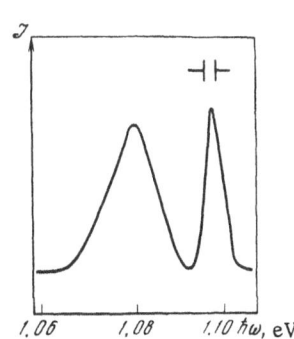

Fig. 32. Photoluminescence spectrum of Si (ρ = 350 $\Omega \cdot$ cm at T = 300°K), obtained at T = 15°K for I = 6×10^{18} cm$^{-2} \cdot$ sec^{-1}.

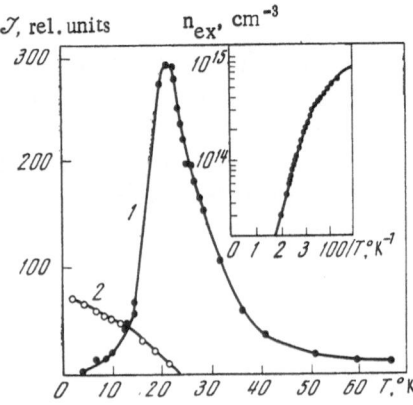

Fig. 33. Temperature dependences of the intensities of the luminescence bands at 1.098 eV (1) and 1.083 eV (2). The inset shows the calculated temperature dependence of the free-exciton density (continuous curve) and the experimentally determined temperature dependence of the integrated free-exciton luminescence intensity (points).

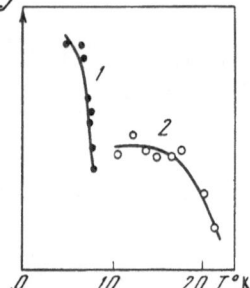

Fig. 34. Temperature dependences of the intensity of the luminescence due to bound (1) and free (2) excitons.

§4. Determination of the Binding Energy of Free Excitons from the Fall of the Luminescence Intensity with Rising Temperature

Figure 33 shows that only the free-exciton luminescence was observed in the photoluminescence spectrum of silicon obtained for $I \approx 6 \times 10^{18}$ cm$^{-2} \cdot$sec^{-1} at T > 22°K. Obviously, in this temperature range the equilibrium is due to the exchange of particles between the exciton system and the electron–hole gas. A temperature rise increases the probability of the thermal dissociation of free excitons and, as a consequence, it reduces the luminescence intensity. In the temperature range where only the exciton annihilation is observed in the absence of fast nonradiative recombination channels, the equilibrium density of excitons can be found from the conditions $N = n_{ex} + n_{e-h}$, where N is the concentration of the optically generated nonequilibrium carrier pairs, n_{ex} is the number of excitons per unit volume, and n_{e-h} is the number of electron–hole pairs (per unit volume) which are not bound into excitons. Obviously, at low optical excitation rates when the concentration of nonequilibrium carriers is far from degeneracy, n_{e-h} and n_{ex} are described by the following dependences [38]:

$$n_{e-h} = 2 \left(\frac{m^*_{e,h} kT}{2\pi\hbar^2} \right)^{3/2} e^{-\frac{|F_{e,h}|}{kT}},$$

$$n_{ex} = g \left(\frac{MkT}{2\pi\hbar^2} \right)^{3/2} e^{-\frac{F-\mathscr{E}_0}{kT}},$$

where $F_{e,h}$ are the quasi-Fermi levels of electrons and holes, $M = m^*_e + m^*_h$, $F = F_e + F_h$, and g is the degree of the spin degeneracy of the exciton state.

In the case of semiconductors with similar effective masses of electrons and holes ($m_e^* \approx m_h^* = m^*$) the concentration of excitons is given by

$$n_{ex} = n_T e^{-\mathscr{E}_0/kT} \left(\frac{1}{2} + \frac{N}{n_T e^{-\mathscr{E}_0/kT}} - \frac{1}{2} \sqrt{1 + \frac{4N}{n_T e^{-\mathscr{E}_0/kT}}} \right), \tag{22}$$

where

$$n_T = \left(\frac{m^* kT}{2\pi\hbar^2} \right)^{3/2}.$$

The quantity $n_T e^{-\mathscr{E}_0/kT}$ represents the number of states in the allowed bands occupied by electrons and holes liberated as a result of thermal dissociation of excitons. At low temperatures, when $n_T e^{-\mathscr{E}_0/kT} \ll N$, practically all the nonequilibrium carriers are bound into excitons. When the opposite inequality is obeyed, the solution of the above equation is an expansion in powers of the small parameter $N/n_T e^{-\mathscr{E}_0/kT}$ In this temperature range the thermal dissociation of excitons reduces their concentration at a nearly exponential rate governed by the binding energy of excitons, i.e.,

$$n_{ex} n_T = N^2 e^{\mathscr{E}_0/kT} \tag{23}$$

A semilogarithmic calculated temperature dependence of the exciton concentration is plotted in Fig. 33 for $N = 10^{15}$ cm^{-3}, $\mathscr{E}_0 = 14$ meV, and $m = m_e^* = 1.08 m_e$. It is clear from Fig. 33 that the experimental results are in good agreement with the curve calculated on the basis of Eq. (22). The replacement of m^* with a value in the range $0.55m < m^* < 1.08m$ simply shifts the calculated curve along the horizontal axis and hardly changes its nature in this range of effective masses. The binding energy of excitons deduced from the fall of the luminescence intensity with rising temperature is close to the binding energy of free excitons in Si obtained from the optical absorption [62] and calculated in [61]. Here, $\mathscr{E}_0$ represents the slope of the calculated curve so that the precision of this quantity is limited by the adopted hypothesis that the effective masses of electrons and holes are equal.

The average concentration of nonequilibrium carriers at the excitation rate employed is only an order-of-magnitude estimate because in the range of validity of Eq. (22) the calculated value of N depends on the selected value of the effective mass m^*.

The temperature dependences of the intensity of the free-exciton luminescence calculated for different optical excitation rates are plotted in Fig. 35. We can see that the temperature of the transition (inflection) from the region characterized by an almost constant exciton concen-

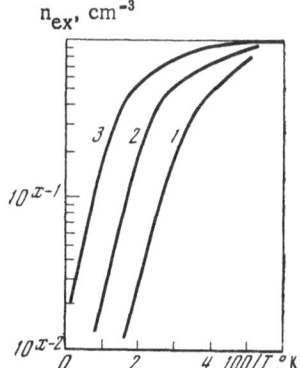

Fig. 35. Calculated temperature dependences of the free-exciton density in the case of steady-state excitation of nonequilibrium carriers of different concentrations (cm^{-3}): 1) 10^{15} ($x = 15$); 2) 10^{16} ($x = 16$); 3) 10^{17} ($x = 17$).

tration to the region of the exponential fall of the luminescence intensity depends strongly on the optical excitation rate.

§ 5. Discussion of Experimental Results

We shall now summarize briefly the experimental results obtained. At low temperatures but depending on the optical excitation rate we find that before free excitons recombine to produce photons they are quite likely either to form a bound state with impurities present in the semiconductor or to assume a new state for which the recombination is characterized by a wide luminescence band at 1.083 eV (Fig. 29). The width of this band is considerably greater than the value expected from the thermal broadening and the luminescence maximum is separated from the corresponding free-exciton luminescence band by about 15 meV (Fig. 32). The binding energy of free excitons in Si is 14 ± 0.5 meV.

According to the first interpretation given by Haynes, this luminescence is due to the formation of exciton molecules — biexcitons — at low temperatures. An exciton molecule does not differ in any basic respect from a free exciton except that it is composed of four particles with a large radius of rotation compared with the corresponding radius of a free exciton. It is shown in Chap. I that, beginning from concentrations $Na^3 \approx 10^{-2}$, the existence of biexcitons is less favorable from the energy point of view than the formation of liquid-phase drops. This follows from the conclusion that at these concentrations it is likely that exciton molecules dissociate into free excitons. Using the calculated results given in § 4 in the present chapter, we can show that when the optical excitation rate exceeds 10^{22} cm^{-2}·sec^{-1}, the photoluminescence spectrum of Si should not have a band or line due to the recombination of exciton molecules.

It follows from Fig. 31 that the new luminescence is observed right up to the highest optical excitation rates. The slow rise of this luminescence and the increase in the intensity of the free-exciton luminescence are largely due to the overheating of a surface layer of the semiconductor, as shown by the experimental results discussed in § 3. Moreover, the energy balance proposed by Haynes for the recombination of exciton molecules is valid only if the binding energy of free excitons is $\mathcal{E}_0 = 8$ meV. This value is in conflict with our binding energy and with the energy deduced in [62] from the low-temperature absorption spectra of Si.

Our explanation of the experimental observations is based on the assumption that a gas—liquid phase transition occurs at high exciton concentrations.

It can be demonstrated that this phenomenon occurs in Si by calculating the energy scheme of interacting electrons and holes at different concentrations, in the same way as it was done for GaAs in Chap. III. Such a calculation for Si is reported in [70]. We shall calculate the exchange energy of the interaction between electrons allowing for the ν minima of the conduction band. Then,

$$E^s_{\text{exch}} = -\frac{0.916}{r_s/a_0}\frac{R}{\chi \nu^{1/3}} \ [\text{eV}], \qquad E^h_{\text{exch}} = -\frac{0.916}{r_s/a_0}\frac{R}{\chi} \ [\text{eV}].$$

The correlation energy of electrons and holes can still be determined using the Wigner interpolation formula (see Chap. II).

There is as yet no analytic expression for the energy of the interaction of carriers of opposite sign E^{e-h}_{corr} in the situation when they are completely screened. However, clearly E^{e-h}_{corr} should not exceed the sum of the correlation energies of electrons and holes. It is clear from Fig. 36 that a system of electrons and holes has a stable minimum at $N = 4 \times 10^{18}$ cm^{-3} and the binding energy of a pair of particles at this minimum is 16 meV. Estimates of the concentrations corresponding to the thermal degeneracy threshold of carriers $N^{e,\,h}_T$ and to the exciton dissociation threshold due to the screening of the Coulomb potential are given in Table 5.

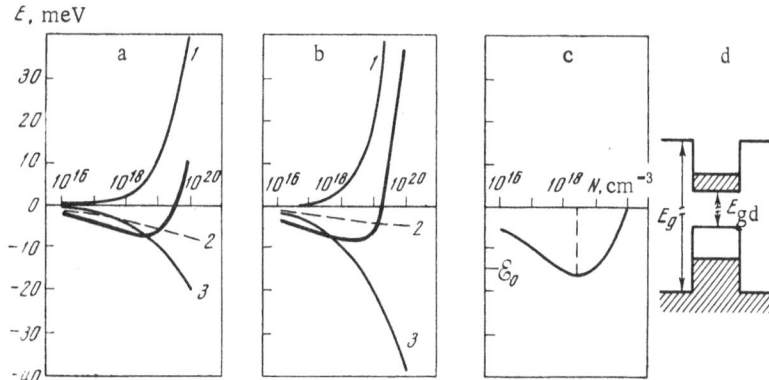

Fig. 36. Energy relationships calculated for electrons and holes interacting in silicon: a) dependence of the energy of interacting electrons on their concentration; b) dependence of the energy of interacting holes on their concentration; c) dependence of the energy per electron – hole pair on their concentration; d) electron – hole condensate model: 1) E_{kin}; 2) E_{corr}; 3) E_{exch}; the thick curves in Figs. 36a and 36b represent the dependences of the energy $E = E_{kin} + E_{corr} + E_{exch}$ on the carrier concentration.

In contrast to GaAs, almost all the nonequilibrium carriers present at low temperatures in Si are in the bound state. For this reason, N_{T-F} does not represent the maximum concentrations of excitons n_{ex}. Nevertheless, all the estimates in Table 5 indicate that at $T \leq 77°K$ for $N = 4 \times 10^{18}$ cm^{-3} a nonequilibrium system of electrons and holes is a strongly degenerate Fermi gas. At low temperatures this gas is absolutely stable in the sense that the kinetic energy of electrons and of holes is less than the potential energy due to their collective interaction (Fig. 36d).

At low values of N there is a discrete level in Fig. 36 and this level represents the binding energy of excitons $\mathcal{E}_0$, whose value is less than the sum of the energies of an electron and a hole for $N = 4 \times 10^{18}$ cm^{-3}, the difference being $E_g - E_{gd} - \mathcal{E}_0 = 2$ meV. The situation is favorable for the appearance of a phase transition of the first kind.

Since the experimental results indicate that the long-wavelength luminescence band appears already for $N \geq 5 \times 10^{14}$ cm^{-3} and the intensity of this band rises continuously with the optical excitation rate, whereas the maximum is hardly shifted along the energy scale, we may conclude that the condensation of excitons begins with the formation of electron–hole drops.

It is clear from Fig. 31 that when the energy of electrons and holes in drops increases by an amount exceeding $E_g - E_{gd} - \mathcal{E}$, strong evaporation from electron–hole drops begins and the equilibrium of the system is shifted in the direction of higher exciton concentrations. According to this model, there should be some critical temperature above which the condensation of an exciton gas does not occur. It is clear from Fig. 36 that the disappearance of the long-

TABLE 5

T, °K	N_T^e, cm^{-3}	N_T^h, cm^{-3}	N_{Deb}, cm^{-3}	N_{T-F}^e, cm^{-3}	N_{T-F}^h, cm^{-3}	Remarks
77	$3.2 \cdot 10^{18}$	$1.2 \cdot 10^{18}$	$4 \cdot 10^{17}$	—	—	N_{Deb} calculated from $L_{Deb} = a = 40$ Å
4.2	$4.1 \cdot 10^{16}$	$3.5 \cdot 10^{16}$	$2 \cdot 10^{16}$	$2.8 \cdot 10^{15}$	$2.2 \cdot 10^{16}$	N_{T-F} calculated from $L_{T-F} = a = 40$ Å

wavelength luminescence occurs at a temperature corresponding to kT $\approx$ 1.9 meV. However, this agreement between the experimental and calculated results should not be regarded as promising because the calculated value of $E_g - E_{gd}$ is underestimated and the thermal drop dissociation threshold should depend on the nonequilibrium carrier concentration.

Nevertheless, all the experimental results presented in this chapter can be explained qualitatively on the basis of the exciton condensate model.

CHAPTER VI

PHOTOELECTRIC PROPERTIES OF SILICON AT HIGH OPTICAL EXCITATION RATES

§1. Review of Literature

The structure of the condensed exciton phase results in a strong "collectivization" of all electrons and holes so that they can move independently of one another. In this sense the dense phase resulting from the condensation of excitons should be similar to a liquid metal. However, since both excitons and metallic exciton drops carry no net charge, they do not contribute to the conductivity of a semiconductor until either the Mott transition of excitons to the metallic state takes place or until the new phase fills a certain volume between the current-carrying contacts.

In the case of a many-electron system of hydrogen-like centers the transition to a state with a metallic conduction is quite sudden [74] because, due to the screening and the collective nature of the process, each liberated electron helps to liberate other electrons. The Mott transitions were first observed [75] in studies of the impurity conduction in Ge and Si. The sudden transition of a system to the metallic state was predicted by Mott to occur when $Na^3 = 0.25$, where N is the concentration of hydrogen-like centers; this condition was confirmed experimentally.

Since the Bohr radius of excitons in semiconductors is considerably greater than the Bohr radius of impurity centers, the metallization of excitons should occur at relatively low concentrations and this should be easy to observe experimentally.

Investigations of this type were first carried out on Ge [76]. It was found that in prebreak-down fields when the nonequilibrium carrier density reached 2×10^{15} cm^{-3} the conductivity suddenly rose. This rise was attributed to a considerable increase in the number of free carriers as a result of a transition of the system to the metallic state. This transition was assumed to be supported by the absence of the temperature dependence of the conductivity beyond the transition point. The concentration at which the metallization of excitons took place satisfied the condition $Na^3 \geq 0.2$, where $a = 140$ Å is the Bohr radius of excitons in Ge.

These investigations stimulated studies of the photoelectric properties of silicon [71, 77]. The experimental results [71] indicated that studies of just the conductivity of semiconductors in the case of strong injection of nonequilibrium carriers were insufficient for the interpretation of the physical processes which occurred under these circumstances. It should be pointed out that nonequilibrium carriers were injected in [77] by a pulsed electric field. The strong rise of the electrical conductivity observed for N > 10^{18} cm^{-3} was explained, as in [76], by the metallization due to the Mott transition of excitons.

§2. Measurement Method

Since all the effects associated with the metallization of Si should occur at nonequilibrium carrier concentrations exceeding considerably the corresponding concentrations in Ge, we used high-power solid-state lasers which produced unfocused photon fluxes up to I $\approx 10^{25}$ cm$^{-2} \cdot$ sec^{-1}.

The circuit used to measure the photoconductivity signal consisted of a silicon sample, a load resistance, and a low-resistance static voltage source, all connected in series. It is easy to show that the amplitude of the voltage across a sample v, due to a modulated illumination of a crystal, should be described by

$$v = V \frac{\Delta \Sigma r_0 R \, (r_0 - \Delta r)^2}{(R + r_0)(R + r_0 - \Delta r)^2},$$

where r_0 is the dark resistance of the sample; V is the constant voltage applied to the contacts; Δr is the change in the resistance of the sample as a result of illumination; $\Delta \Sigma$ is the corresponding change in the conductance; R is the load resistance. Obviously, if $R \gg r_0 - \Delta r$, we should have

$$v \propto V \frac{\Delta r}{R(R + r_0)},$$

and we should obtain $v \propto VR\Delta\Sigma$ only for $R \ll r_0 - \Delta r$. Thus, the ohmic load should have a very low resistance in order to ensure that the measured voltage reflected the change in the conductance of the sample due to illumination at all illumination intensities. This resistance was estimated for each specific case. At high excitation levels the value of R was a fraction of an ohm.

The current contacts were prepared by the evaporation of Ag on the opposite faces of a sample and this was followed by annealing at $T \approx 700°K$ in high vacuum. The samples prepared in this way were characterized by negligible values of the photo-emf at the metal−semiconductor contacts.

§3. Photoluminescence Spectra of Si in the Presence

of Static Electric Fields. Impact Ionization of Free

Excitons

At low temperatures a considerable redistribution of the intensities between the various radiative recombination processes occurred when the static field applied to a sample was increased. Qualitatively, the variations of the intensity of the long-wavelength luminescence band at 1.083 eV and of the bands due to bound and free excitons were similar to the variations observed in the luminescence spectra when the temperature was increased. Figures 37 and 38 show, for the sake of comparison, the photoluminescence spectra of Si obtained at different temperatures exceeding 4.2°K (Fig. 37) and in different electric fields (Fig. 38). In both cases an increase of the external perturbation reduced the intensity of the luminescence due to electron−hole drops and due to bound excitons.

At the critical point corresponding to the total disappearance of the 1.083 eV energy the intensity of the free-exciton band $\mathscr{I}_{ex}$ reached its maximum value. However, in an electric field the absolute value of this intensity $\mathscr{I}_{ex}$ was less than that observed on thermal dissociation of electron−hole drops. Moreover, in a "strong" electric field the free-exciton luminescence maximum was shifted in the direction of higher energies by 2 meV and a new luminescence band, separated by $\Delta\hbar\omega = 13$ meV, appeared in the short-wavelength wing.

Similar characteristic changes in the photoluminescence spectra were observed at low optical excitation rates when electron−hole drops were not formed. In this case the majority of free carriers was found to be bound to boron atoms and only one luminescence line at 1.092 eV was observed in the spectra (Fig. 30). When the temperature was increased, the thermal

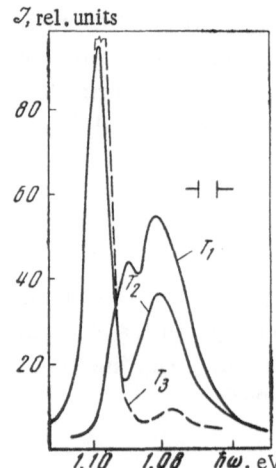

Fig. 37. Photoluminescence spectra of Si recorded for
$I = 5 \times 10^{18}$ cm^{-2} · sec^{-1} at three different temperatures
$T_1 < T_2 < T_3$.

ionization of bound excitons was accompanied by the appearance of the free-exciton luminescence (Fig. 34).

When the electric field was increased, the bound-exciton luminescence intensity decreased
continuously exhibiting a particularly strong fall in fields E > 20 V/cm (Fig. 39). No free-exciton
luminescence was observed within the limits set by the sensitivity of the recording system.

A qualitative comparison of the photoluminescence spectra disturbed by an external elec-
tric-field heating indicated that the reduction in the intensity of the electron−hole and bound-
luminescence was observed within the limits set by the sensitivity of the recording system.
sult of dissociation of free excitons. It is clear from Fig. 38 that the changes in the photoluminescence spectra did not have a threshold. The ionization of excitons was supported by
measurements of the dependence of the photocurrent flowing through a sample in which nonequilibrium carriers were generated by short light pulses on the applied electric field. A
current−voltage characteristic of this type is shown in Fig. 40: this characteristic was obtained for $I \approx 10^{21}$ cm^{-2} · sec^{-1} and it had a region of a strong rise in the current indicating
the existence of exciton breakdown fields.

The ionization threshold of excitons in an electric field could not be determined accurately.
The breakdown fields depended on the excitation rate and, in contrast to the impurity breakdown
[78], the ionization of excitons was manifested by a relatively stronger rise of the photoconductivity signal in fields exceeding a certain threshold value.

A characteristic feature of the exciton ionization, which distinguishes it from the impurity
breakdown, is that only light particles of mass close to that of a free electron participate in
this process. Moreover, an exciton is a weakly bound entity with an ionization energy lower
than the ionization energy of impurities. The interaction of free carriers with excitons is

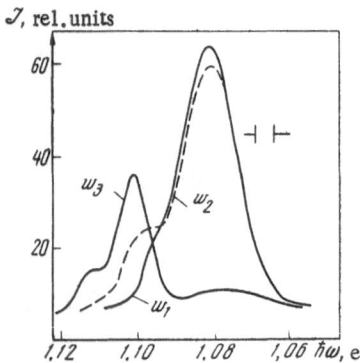

Fig. 38. Photoluminescence spectra of Si recorded for
$I = 8 \times 10^{18}$ cm^{-2} · sec^{-1} in different electric fields: w_1 =
0; w_2 = 30 V/cm; w_3 = 40 V/cm.

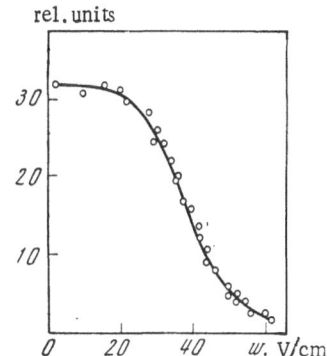

Fig. 39. Dependence of the intensity of the bound-exciton luminescence band at E = 1.092 eV on the electric field (T = 4.2°K).

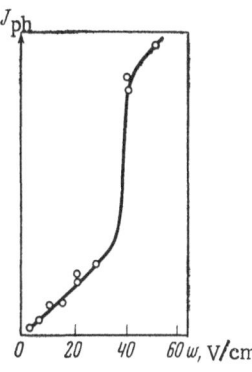

Fig. 40. Dependence of the photocurrent on the electric field obtained for I ≈ 10^{21} cm^{-2} · sec^{-1} at 4.2°K.

inelastic even in fields below the breakdown value. For these reasons the exciton breakdown is not as abrupt as the impurity breakdown.

Since electrostatic ionization of excitons would require fields many times higher than those found experimentally, the most probable exciton ionization mechanism is the impact process, exactly as in the case of impurities. This ionization mechanism of free excitons is supported also by the experimental results given above.

Further support of this interpretation is provided by the appearance of the interband luminescence in "strong" electric fields. The separation between the maxima of the luminescence due to excitons and due to free electrons and holes is $\Delta\hbar\omega \approx 13$ meV. The value of $\Delta\hbar\omega$ is close to the binding energy of free excitons determined in § 4, Chap. V. However, it should be noted that since the interband luminescence band is not sufficiently well resolved in the short-wavelength wing of the free-exciton band, the binding energy of free excitons found in this way may be underestimated by 1-2 meV.

The influence of a static electric field on the photoluminescence spectra of Ge and Si was recently considered in [79]. The experimental results were explained by the Haynes model of exciton molecules and a reduction in the intensity of the long-wavelength luminescence was attributed in [79] to the dissociation of these molecules in an electric field. The results of the investigation reported above exclude the possibility of this interpretation. Moreover, it should be noted that in the case of n-type samples used in [79] a very strong luminescence line of free excitons at $\omega = 0$ is evidence of a considerable overheating of the semiconductor by the high-power optical excitation.

§ 4. Kinetics of Recombination Processes in Si

Figure 41 shows an oscillogram of a photocurrent pulse generated by the excitation of silicon with ruby laser radiation. This oscillogram was obtained at T = 4.2°K at an optical excitation level which ensured that the luminescence was dominated by electron−hole drops. The duration of the photoresponse pulse, determined at its mid-amplitude, was $\tau = 50$ nsec.

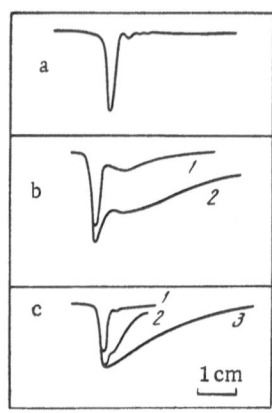

Fig. 41. Oscillograms of photocurrent pulses in silicon under different conditions: a) $T = 4.2°K$, $I = 10^{21}$ cm^{-2} · sec^{-1} (1 cm = 250 nsec); b) $T = 8°K$, $I = 10^{22}$ cm^{-2} · sec^{-1} (curves 1 and 2 correspond to 1.8 and 40 V/cm, respectively); c) $T = 8°K$; $w = 2$ V/cm (curves 1, 2, and 3 correspond to $I = 2 \times 10^{21}$, 5×10^{22}, and 10^{23} cm^{-2} · sec^{-1}). The time scale in Figs. 41b and 41c is 1 cm = 1 μsec.

In fields close to the breakdown value and under strong photoexcitation conditions, when the heating of the semiconductor became important, the time characteristics showed relaxation with a time constant τ exceeding 2 μsec, which was governed by the nonequilibrium carrier and exciton lifetimes in Si at liquid-helium temperatures.

§5. Investigation of Excitons at High Concentrations

in Weak Electric Fields

The purpose of the investigation described below was to determine the nature of the changes in the conductivity of silicon due to the transition of the exciton system from the dielectric to the metallic state.

The photoelectric properties were measured in fields known to be below the breakdown value. Figure 42 shows the dependences of the photocurrent on the optical excitation rate obtained in different electric fields at $T = 4.2°K$. In a field $w < 10$ V/cm this dependence was step-like. The concentration of the carriers N generated by the absorption of light, plotted along the abscissa, was determined from the room-temperature photoconductivity when the electron and hole mobilities were known.

It is clear from Fig. 42 that the rise of the photoconductivity signal occurred up to a certain concentration of nonequilibrium carriers (4×10^{17} cm^{-3}) and this was followed by the saturation of the photocurrent in the range 4×10^{17} cm$^{-3} \leq N \leq 2 \times 10^{18}$ cm^{-3}. A second rise of the photocurrent J_{ph} was observed at $N \approx 2 \times 10^{18}$ cm^{-3} and this was followed by a saturation region. This last saturation region of J_{ph} was observed at the optical excitation rates such that the photoresistance of the sample during illumination became close to the resistance of the ohmic load.

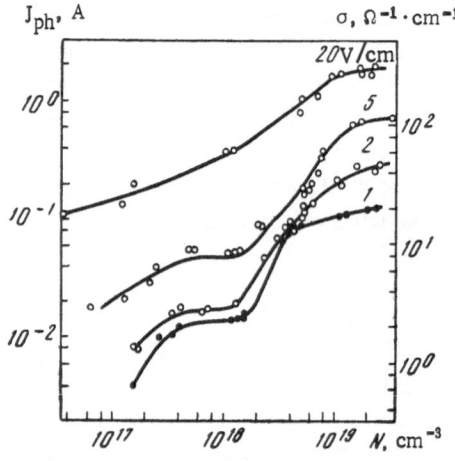

Fig. 42. Dependences of the photocurrent on the carrier concentration in silicon at 4.2°K in different electric fields (the values are given alongside the curves). The scale on the right-hand side gives the conductivity calculated for the curve recorded using a field $w = 1$ V/cm.

Characteristic features of the dependence $\sigma(N)$ observed in weak electric fields at helium temperatures can be understood by postulating the appearance of electron–hole drops in silicon. Then, the initial rise of $\sigma(N)$ which occurs at $N = 10^{17}$-10^{18} cm^{-3} and is followed by the photocurrent saturation region can be explained by the contribution of nondegenerate electrons and holes which are not bound into drops. In the range of N where $\sigma(N)$ = const, an increase in the density of free carriers is compensated by a reduction in their mobility.

According to the results reported in Chap. V a strong rise of the photocurrent in the $N > 2 \times 10^{18}$ cm^{-3} range can be explained by the metallization of the surface layer between the current-carrying contacts. The nature of the dependence of σ on N should be governed by a new type of conduction due to the transition of the carrier system to a degenerate state. It is shown in [76] that the suppression of the electron–hole scattering in strongly degenerate Ge gives rise to an experimental dependence of the conductivity which is proportional to N^3.

However, since in Si all the effects associated with the metallization of this semiconductor occur at high optical excitation rates, the heating effect is the dominant factor in the dependence of σ on N. The average temperature rise in the active region is estimated to be $\overline{\Delta T} \approx 10°K$. It is in this range, which corresponds to $N > 10^{18}$ cm^{-3}, that the intensity of the free-exciton luminescence rises strongly (Fig. 31) and the time characteristics of the photoconductivity become broader (Fig. 41).

Since the new luminescence appears in the spectra for $N \geq 10^{14}$ cm^{-3} and its intensity rises continuously right up to the highest optical excitation rates and since the luminescence maximum hardly shifts along the energy scale, serious difficulties are encountered if the experimental results are attributed to the dissociation of exciton molecules or the metallization of the semiconductor as a result of the Mott transition.

The faster rise of the electrical conductivity of Si observed at $N \approx 2 \times 10^{18}$ cm^{-3} [77] and attributed to the Mott transition of excitons to a state with the metallic type of conduction may be due to a process similar to that discussed above.

The author is grateful to his scientific director V. S. Bagaev whose constant help in the experiments and guidance have helped greatly in the completion of this investigation. The author is greatly indebted to L. V. Keldysh for numerous helpful discussions of the results and valuable advice. Thanks are due to the Head of the Laboratory of Semiconductor Physics Academician B. M. Vul for discussing the study and for critical comments, and to the Leader of the Section of Photoelectric Processes A. P. Shotov for providing the facilities needed to investigate the selected topic and for his valuable comments. This opportunity is also taken to express thanks to B. D. Kopylovskii for his constant help on the methodological part of the study, and to O. V. Gogolin, V. B. Stopachinskii, and all staff of the Laboratory who helped in one way or another. The author is grateful to Ya. E. Pokrovskii and A. É. Yunovich for discussing the results and critical comments.

LITERATURE CITED

1. S. A. Moskalenko, Fiz. Tverd. Tela, 4:276 (1962); R. C. Casella, J. Appl. Phys., 34:1703 (1963).

2. L. D. Landau and E. M. Lifshitz, Statistical Physics, 2nd ed., Pergamon Press, Oxford (1969).

3. L. V. Keldysh, Proc. Ninth Intern. Conf. on Physics of Semiconductors, Moscow, 1968, Vol. 2, publ. by Nauka, Leningrad (1969), p. 1303; Collection: Excitons in Semiconductors [in Russian], Nauka, Moscow (1971), p. 5.

4. L. V. Keldysh and A. N. Kozlov, Zh. Éksp. Teor. Fiz., 54:978 (1968).

5. M. A. Lampert, Phys. Rev. Lett., 1:450 (1958).

6. S. A. Moskalenko, Opt. Spektrosk., 5:147 (1958).
7. R. R. Sharma, Phys. Rev., 170:770 (1968).
8. A. M. Bonch-Bruevich, Radio Engineering in Experimental Physics [in Russian], Nauka, Moscow (1966).
9. R. A. Smith, F. E. Jones, and R. P. Chasmar, Detection and Measurement of Infrared Radiation, 2nd ed., Clarendon Press, Oxford (1957).
10. J. Strong et al., Procedures in Experimental Physics, Prentice-Hall, New York (1938).
11. V. S. Vavilov, A. F. Plotnikov, and B. D. Kopylovskii, Prib. Tekh. Éksp., No. 3, 183 (1962).
12. B. G. Cohen, W. B. Snow, and A. R. Tretola, Rev. Sci. Instrum., 34:1091 (1963).
13. V. I. Osinskii and N. N. Sirota, Vestn. Akad. Navuk B. SSR Ser. Fiz.-Mat. Navuk, No. 3, 130 (1965).
14. M. G. Holland, Phys. Rev., 134:A471 (1964).
15. É. I. Zavaritskaya, Tr. Fiz. Inst. Akad. Nauk SSSR, 37:41 (1966).
16. R. W. Keyes, IBM J. Res. Dev., 9:303 (1965).
17. V. I. Shveikin and L. I. Paduchikh, Radiotekh. Élektron., 13:101 (1968).
18. G. J. Lasher and W. V. Smith, IBM J. Res. Dev., 8:532 (1964).
19. Y. Toyozawa, Prog. Theor. Phys. Suppl., No. 12, 111 (1959).
20. M. D. Sturge, Phys. Rev., 127:768 (1962).
21. R. J. Elliot, Phys. Rev., 108:1384 (1957).
22. M. I. Nathan and G. Burns, Phys. Rev., 129:125 (1963).
23. C. Benoit à la Guillaume and C. Tric, J. Phys. Chem. Solids, 25:837 (1964).
24. R. R. Sharma and S. Rodriguez, Phys. Rev., 153:823 (1967).
25. R. R. Sharma and S. Rodriguez, Phys. Rev., 159:649 (1967).
26. J. J. Hopfield, Proc. Seventh Intern. Conf. on Physics of Semiconductors, Paris, 1964, Vol. 1, Physics of Semiconductors, publ. by Dunod, Paris; Academic Press, New York (1964), p. 725.
27. E. H. Bogardus and H. B. Bebb, Phys. Rev., 176:993 (1968).
28. M. A. Gilleo, P. T. Bailey, and D. E. Hill, Phys. Rev., 174:898 (1968); R. C. C. Leite, J. Shah, and J. P. Gordon, Phys. Rev., Lett., 23:1332 (1969).
29. V. S. Bagaev and L. I. Paduchikh, Fiz. Tverd. Tela, 11:2268 (1969).
30. J. Shah, R. C. C. Leite, and J. P. Gordon, Phys. Rev., 176:938 (1968).
31. Y. Toyozawa, Prog. Theor. Phys., 27:89 (1962).
32. A. I. Ansel'm and Yu. A. Firsov, Zh. Éksp. Teor. Fiz., 30:719 (1956).
33. É. I. Rashba and G. É. Gurgenishvili, Fiz. Tverd. Tela, 4:1029 (1962).
34. J. S. Blakemore, Semiconductor Statistics, Pergamon Press, Oxford (1962).
35. D. B. Wittry and D. F. Kyser, Proc. Eight Intern. Conf. on Physics of Semiconductors, Kyoto, 1966, in: J. Phys. Soc. Jap., Suppl., 21:312 (1966).
36. D. Pines, Elementary Excitations in Solids, Benjamin, New York (1963).
37. W. P. Dumke, Phys. Rev., 132:1998 (1963).
38. W. F. Brinkman, T. M. Rice, P. W. Anderson, and S. T. Chui, Phys. Rev. Lett., 28:961 (1972); M. Combescot and P. Nozieres, Solid State Commun., 10:301 (1972).
39. O. Akimoto and E. Hanamura, Solid State Commun., 10:253 (1972).
40. J. Adamowski and S. Bednarek, Solid State Commun., 9:2037 (1971).
41. Ya. E. Pokrovskii and K. I. Svistunova, ZhÉTF Pis'ma Red., 13:297 (1971).
42. N. N. Sibel'din, V. S. Bagaev, V. A. Tsvetkov, and N. A. Penin, Fiz. Tverd. Tela, 15:177 (1973).
43. V. S. Bagaev, T. I. Galkina, N. A. Penin, V. B. Stopachinskii, and M. N. Churaeva, ZhÉTF Pis'ma Red., 16:120 (1972).
44. V. S. Bagaev, L. I. Paduchikh, and V. B. Stopachinskii, ZhÉTF Pis'ma Red., 15:508 (1972).
45. C. J. Hwang and L. R. Dawson, Solid State Commun., 10:443 (1972).
46. J. D. Cuthbert, J. Lumin., 1:307 (1970).
47. J. A. Rossi, C. M. Wolfe, G. E. Stillman, and J. O. Dimmock, Solid State Commun., 8:2021 (1970).

48.   D. D. Sell, R. Dingle, S. E. Stokowski, and J. V. DiLorenzo, Phys. Rev. Lett., 27:1644 (1971).

49.   D. Bimberg and W. Schairer, Phys. Rev. Let., 28:442 (1972).

50.   A. Baldereschi and N. O. Lipari, Phys. Rev. B, 3:439 (1971).

51.   D. J. Oliver, Phys. Rev., 127:1045 (1962).

52.   N. G. Basov, A. Z. Grasyuk, and V. A. Katulin, Dokl. Akad. Nauk SSSR, 161:1306 (1965).

53.   N. G. Basov, O. V. Bogdankevich, V. A. Goncharov, B. M. Lavrushin, and V. Yu. Sudzilovskii, Dokl. Akad. Nauk SSSR, 168:1283 (1966).

54.   N. G. Basov, A. Z. Grasyuk, V. F. Efimkov, and V. A. Katulin, Fiz. Tverd. Tela, 9:88 (1967).

55.   A. A. Vvedenov, in: Problems in Plasma Theory [in Russian], Gosatomizdat, Moscow (1963), p. 273.

56.   V. S. Bagaev and L. I. Paduchikh, Fiz. Tverd. Tela, 11:3304 (1969).

57.   G. G. Macfarlane, T. P. McLean, J. E. Quarrington, and V. Roberts, Phys. Rev., 108:1377 (1957); Phys. Rev., 111:1245 (1958); J. Phys. Chem. Solids, 8:388 (1959).

58.   G. H. Wannier, Phys. Rev., 52:191 (1937).

59.   G. Dresselhaus, J. Phys. Chem. Solids, 1:14 (1956).

60.   T. P. McLean and R. Loudon, J. Phys. Chem. Solids, 13:1 (1960).

61.   T. P. McLean, Prog. Semicond., 5:53 (1960).

62.   K. L. Shaklee and R. E. Nahory, Phys. Rev. Lett., 24:942 (1970).

63.   J. R. Haynes, M. Lax, and W. F. Flood, Proc. Fifth Intern. Conf. on Semiconductors, Prague, 1960, publ. by Academic Press, New York (1961), p. 423.

64.   J. R. Haynes, Phys. Rev. Lett., 17:860 (1966).

65.   P. J. Dean, J. R. Haynes, and W. F. Flood, Phys. Rev., 161:711 (1967).

66.   W. Van Roosbroeck and W. Shockley, Phys. Rev., 94:1558 (1954).

67.   P. J. Dean, Y. Yafet, and J. R. Haynes, Phys. Rev., 184:837 (1969).

68.   C. Benoit à la Guillaume, F. Salvan, and M. Voos, J. Lumin., 1:315 (1970).

69.   J. D. Cuthbert, J. Lumin., 1:307 (1970).

70.   A. S. Kaminskii and Ya. E. Pokrovskii, ZhÉTF Pis'ma Red., 11:381 (1970); A. S. Kaminskii, Ya. E. Pokrovskii, and N. V. Alkeev, Zh. Éksp. Teor. Fiz., 59:1937 (1970).

71.   V. S. Bagaev and L. I. Paduchikh, in: Excitons in Semiconductors [in Russian], Nauka (1971), p. 50; Fiz. Tverd. Tela, 13:484 (1971).

72.   B. M. Ashkinadze, I. P. Kretsu, A. A. Patrik, and I. D. Yaroshetskii, Fiz. Tekh. Poluprovodn., 4:2206 (1970).

73.   B. N. Brockhouse, Phys. Rev. Lett., 2:256 (1959).

74.   N. F. Mott, Philos. Mag., 6:287 (1961).

75.   N. F. Mott and W. D. Twose, Advan. Phys., 10:107 (1961).

76.   V. M. Asnin, A. A. Rogachev, and S. M. Ryvkin, Fiz. Tekh. Poluprovodn., 1:1742 (1967); 1:1740 (1967).

77.   V. N. Antonov, Yu. V. Kopaev, Yu. I. Pashintsev, and A. V. Rakov, Fiz. Tverd. Tela, 11:1422 (1969).

78.   I. V. Kucherenko, Thesis for Candidate's Degree, Lebedev Physics Institute, Academy of Sciences of the USSR, Moscow (1969).

79.   A. R. Hartman and R. H. Rediker, Proc. Tenth Intern. Conf. on Physics of Semiconductors, Cambridge, Mass., 1970, publ. by US Atomic Energy Commission, Washington, DC (1970), p. 202.